Published by BlueStrike

Written and designed by Sylvain Richer de Forges

Copyright @ 2019, Sylvain Richer de Forges

All rights reserved. No part of this publication may be reproduced, stored in a retrieval system or transmitted, in any form, or by any means, electronic, mechanical, photocopying recording or otherwise, without the prior written permission of the copyright holder.

Pictures and Diagrams: all diagrams and pictures were used with authorization or from sources not subject to copyright.

Visit the program: www.bluestrike-group.com

Climate change is one of the most serious threat to humanity and should be taken very seriously...

The field of climate research is very dynamic and constantly evolving. For the latest updates on Climate Change research please refer to the regularly updated IPCC reports.

Foreword

The aim of this book is to provide the reader a simple yet complete understanding of Climate Change. Considering the amount of information widely available on the subject, it is easy to get lost, confused or simply overwhelmed with information. Understanding the issue requires an holistic overview of the complex science and impacts of the change. In this view, the author has digested the reliable information available and selected content that clearly highlights the state of knowledge, problems and solutions.

The content of this book is based on scientific facts obtained from most trusted sources mostly from NASA, the IPCC and published scientific papers from respected peered reviewed journals.

The format of this book is unique, rather than exposing the information through lengthy texts, it purposely adopts a series of short key points and graphical representations with sometimes more in-depth explanations. In this way the reader can quickly navigate through complexity in the most simple way and jump between sections without having to read all the book. It is however advised to fully read the entire book as the flow of information is logically an progressively orchestrated. It goes straight to the point of what needs to be known and understood. It clarifies difficult to understand concepts to non scientific readers.

This book should be used by people of all ages and background willing to clearly understand climate change. It primarily targets policy makers and business leaders in a position to make a change.

What is global warming and should I believe it?

I will start this book with the most simple explanation and answer to this fundamental question still on the mind of far too many people. Despite all the controversies surrounding climate change and whether or not it is occurring, debates on the accuracy of the science or uncertainties of measurements, it all comes down to a very simple and undeniable concept: the Greenhouse Effect.

The basics: It can be demonstrated in the laboratory that certain gases have the ability to trap heat when exposed to sunlight. This is not debatable, it is fundamental physics. If you trust the principal behind how your mobile phone operates or other common consumables in modern society, you should also trust the greenhouse effect and we know precisely the warming potential that the different greenhouse gases such as carbon dioxide or methane have.

The problem: We are adding tremendous amounts of greenhouse gases to the atmosphere (a finite space) by burning fossil fuels and through the greenhouse effect this addition undoubtably warms the earth. Whether we like it or not, it is a mathematical certainty.

The best advise I can give to anyone who wants to demystify whether to believe climate change would be to first fully understand the greenhouse effect, all the rest is secondary as they are consequences of this simple phenomenon...

Table of content

I. The Climate System

I.1. Overview of The Climate System
I.2. Heat Transfer
I.3. The Global Energy Balance
I.4. The Global Water Cylce
I.5. The Global Carbon Cycle
I.6. Vulnerability of The Climate System

II. What Is Climate Change?

II.1 What Is Climate Change ?
II.2 Natural Variability
II.3 Examples of Past Climates
II.4 Brief History of Present Climate Change

III. Causes of Climate Change

III.1 Natural Sources of Past Climate Changes
III.2 The Green House Effect
III.3 Green House Gases
III.4 Deforestation

IV. Proofs of Climate Change

IV.1 A Global View
IV.2 Paleoclimatology
IV.3 The CO2-temperature Correlation
IV.4 The Atmospheric CO2 Peak Anomaly
IV.5 The Global Surface Temperature Anomaly
IV.6 The Computer Model
IV.7 Sea Level Rise
IV.8 Ice Mass
IV.9 Ocean Salinity Variations
IV.10 Atmospheric Water Vapor
IV.11 Various Other Proxies

V. Predictions on Climate Change

V.1 Modeling The Climate
V.2 Accuracy of Computer Model
V.3 GHG Predictions
V.4 Sea Level Rise Predictions
V.5 Surface Temperature Predictions
V.6 Precipitations Predictions

VI) Impacts of Climate Change

VI.1 Impacts on Biodiversity
VI.2 Impacts on Ecosystems
VI.3 Impacts on Oceans
VI.4 Impacts on Water Resources
VI.5 Impacts on Societies
VI.6 Impacts on Human Health
VI.7 Impacts on Extreme Weather
VI.8 Impacts on Global Economy

VII) Uncertainties and Ongoing Research on the Impacts of Climate Change

VII.1 Feedback Effects
VII.2 Ice Dynamics
VII.3 Oceanic Circulation
VII.4 Ozone and Climate Change
VII.5 Land Use
VII.6 Global Dimming

VIII) Climate Change in the Singapore Context

VIII.1 The Singapore Context
VIII2 Singapore Emissions and Energy Use
VIII.3 Impacts of Climate Change in Singapore
VIII.4 Singapore Water Resources
VIII.5 Singapore Vulnerability to Sea Level Rise
VIII.6 Health Impacts
VIII.7 Impacts on Singapore's Biodiversity and Ecosystems
VIII.8 Singapore National Climate Change Strategy

IX) What Can Be Done

IX.1 Where are Actions Needed?
IX.2 What can Individuals do?
IX.3 What can Corporations do?
IX.4 What can Governments do?
IX.5 Sustainable Energy
IX.6 Sustainable Development
IX.7 Earth Engineering

X) Common Misunderstandings About Climate Change

Annex 1: Definitions
Annex 2: Acronyms and Abbreviations
Annex 3: Climate Change Quotes
Annex 4: References
Annex 5: Bibliography
Annex 6: Aknowledgements

I. THE CLIMATE SYSTEM

The Climate System: a brief summary

The climate system is governed by numerous interactions between closed loop systems. A modification of one aspect of these systems, such as the carbon cycle, can influence the overall climate system and lead to change...

- **Weather:** Reffers to the changes taking place in the atmosphere at present over short periods of time (the science of meteorology)

- **Climate:** Defines the tendency of the weather over long periods of time (the science of climatology)

The major factors that determine the patterns of climate on earth can be explained in terms of:

- the spherical shape of the earth and the orientation of its axis;

- the strength of the incident radiation from the sun, which determines the overall planetary temperature of the earth;

- the greenhouse effect of water vapour and other radiatively active trace gases;

- the various physical, chemical and biological processes that take place within the atmosphere-geosphere-biosphere climate system, in particular:

- the global energy balance,
- the global water cycle,
- the global carbon cycle and other biogeochemical cycles;

- the rotation of the earth, which substantially modifies the large-scale thermally-driven circulation patterns of the atmosphere and ocean; and

- the distribution of continents and oceans.

The Global Climate System:

- The climate system is driven by a close interaction between the Atmosphere, the Geosphere (Lithosphere, Hydrosphere, Cryosphere) and space

- Human activities can affect/modify these interactions

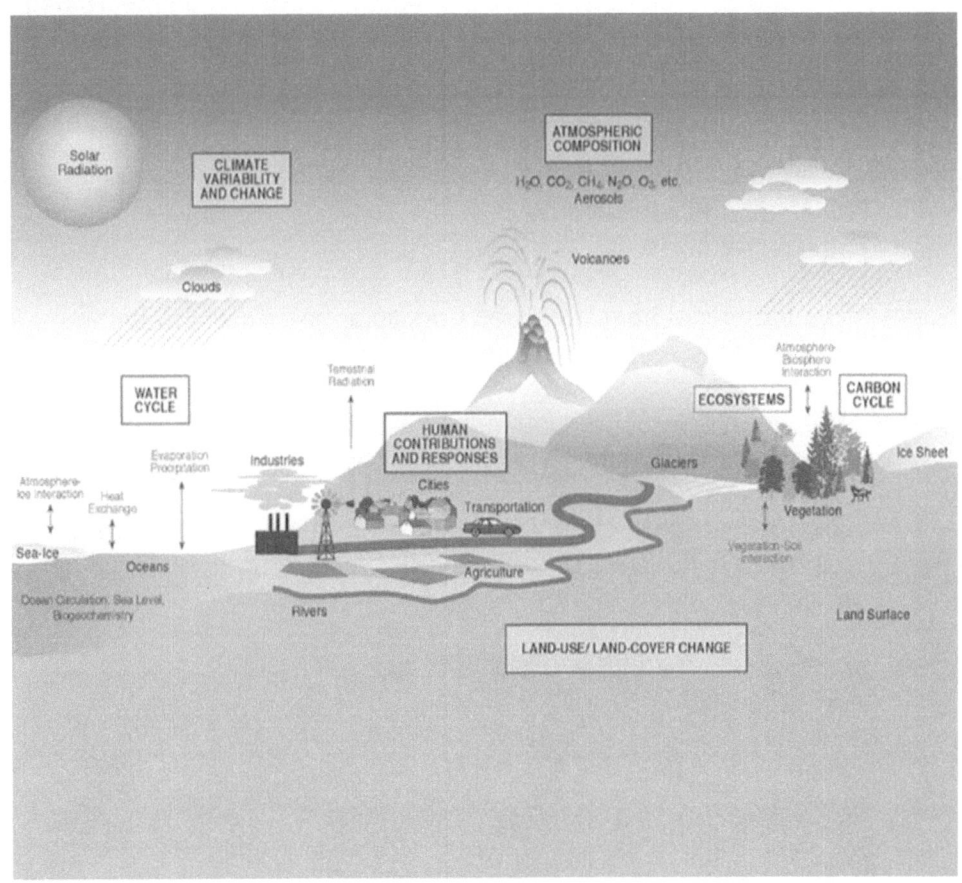

The Climate System: Heat Transfer

- There are constant heat exchanges occuring on the planet between the cold poles and the warm equatorial latitudes

- The earth receives energy from the sun equal to approximately 1360 W m^{-2} (as an annual average) and has an approximate albedo (reflectivity) of 0.3. As such, without any atmosphere the average temperature of the earth would be approximatly – 18 deg C

- There are two main systems by which temperatures are distributed around the planet which affect the weather:

  -The atmospheric air circulations
  -The oceanic currents

- The transfer of heat is derived from solar energy and from the motion of the earth around its axis

- The global temperature of the planet and how it is distributed has an important impact on the behaviour of the climate system

- Climate change is changing this temperature distribution which is expected to have major repercutions on our climate and weather system

Atmospheric circulation transfers a small proportion of the heat around the planet. Oceanic circulation (currents) is by far the main medium for heat transfer around the planet. The general oceanic circulation is known as the oceanic conveyor belt or the thermohaline system

- The global energy balance is the balance between incoming energy from the sun and outgoing heat from the earth

- The global energy balance regulates the state of the earth's climate and modifications to it, as a result of natural and man-made climate-forcing, is causing the global climate to change

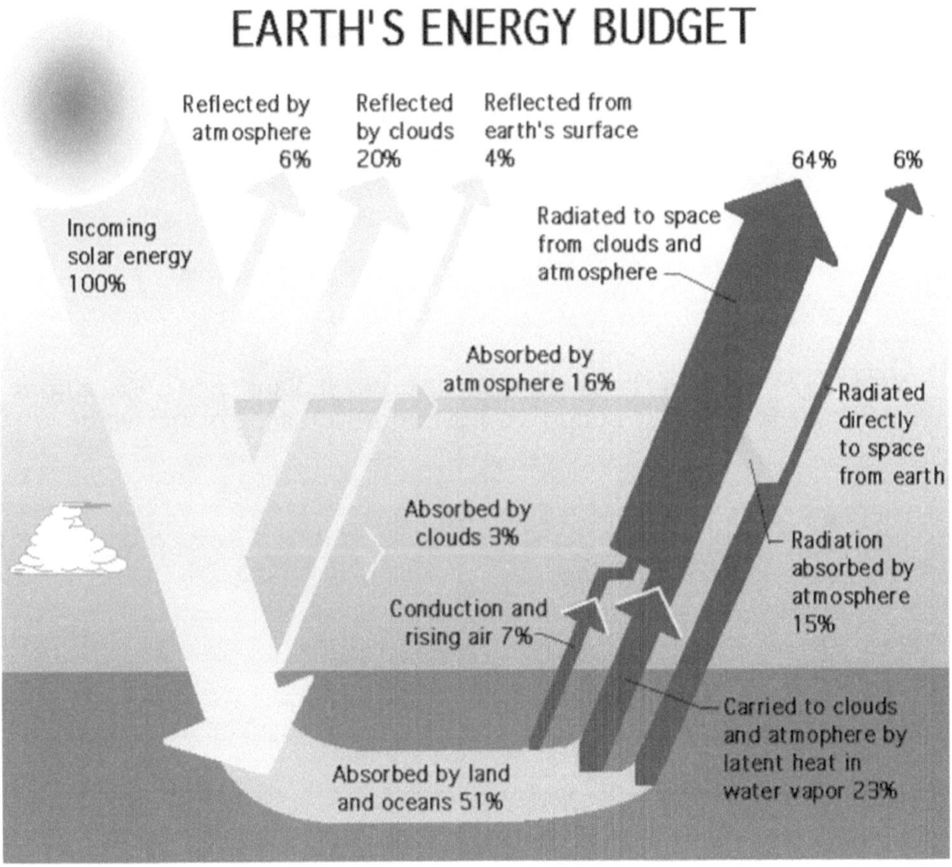

EARTH'S ENERGY BUDGET

Reflected by atmosphere 6%
Reflected by clouds 20%
Reflected from earth's surface 4%

Incoming solar energy 100%

Radiated to space from clouds and atmosphere 64%

Radiated directly to space from earth 6%

Absorbed by atmosphere 16%

Absorbed by clouds 3%

Radiation absorbed by atmosphere 15%

Conduction and rising air 7%

Absorbed by land and oceans 51%

Carried to clouds and atmosphere by latent heat in water vapor 23%

The Climate System: The Global Water Cycle

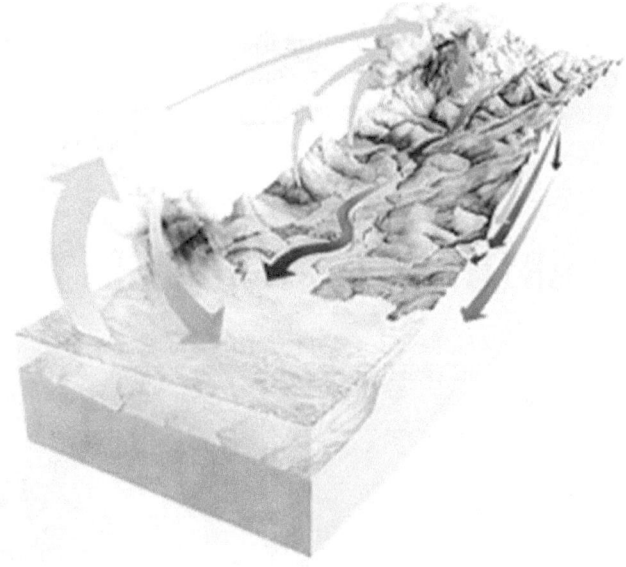

Water (H_2O) is stored in different phases on the planet and passes from one phase to another through a cycle refered to as the global water cycle.

There are 16 main elements of the water cycle:

- *Water storage in oceans*
- *Evaporation*
- *Sublimation*
- *Evapotranspiration*
- *Water in the atmosphere*
- *Condensation*
- *Precipitation*
- *Water storage in ice and snow*
- *Snowmelt runoff to streams*
- *Surface runoff*
- *Streamflow*
- *Freshwater storage*
- *Infiltration*
- *Ground-water storage*
- *Ground-water discharge*
- *Springs*
- *Global water distribution*

- The water cycle plays an important role in regulating the earth climate

- Vise versa, a change in the climate system will significantly influence and modify the water cycle and how water is distributed on the planet

- The transport of atmospheric moisture from the oceans, which cover more than two-thirds of the globe, to the continents plays a vital role to balance the discharge from rivers and groundwater to the oceans

- Water vapour is the most important of the greenhouse gases, in terms of its influence on the climate

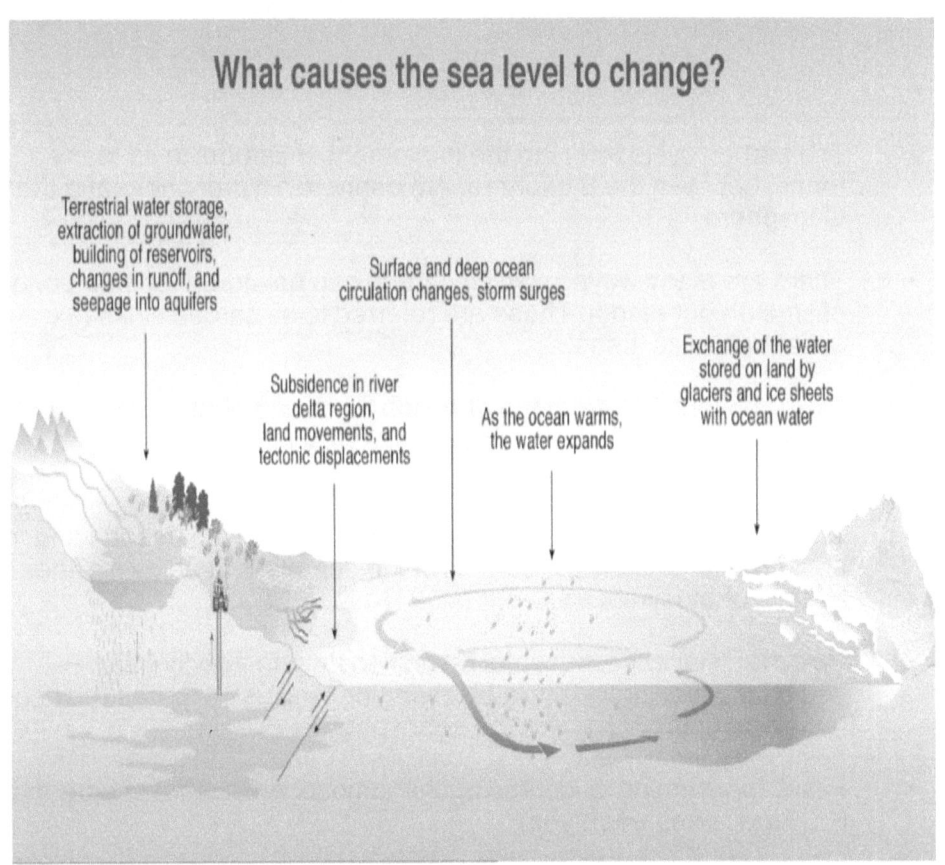

The Climate System: The Global Carbon Cycle

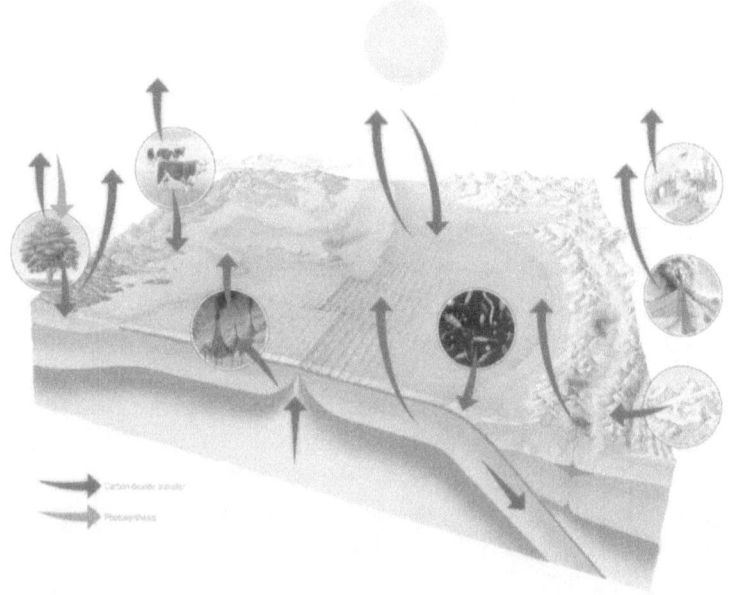

- The carbon cycle refers to the movement of carbon, in its many forms, between the Biosphere, Atmosphere, Hydrosphere and the Geosphere

- There are many ways in which carbon can be stored for long periods of time on our planet. These are referred to as carbon sinks (ex: oceans, forests, trees...)

- Plants absorb CO_2 from the atmosphere during photosynthesis, also called primary production, and release CO_2 back into the atmosphere during respiration

- Another major exchange of CO_2 occurs between the oceans and the atmosphere. The dissolved CO_2 in the oceans is used by marine biota in photosynthesis

- However, since the industrial revolution human activity has influenced the carbon cycle by changing land use and has released huge amounts of CO_2 into the atmosphere

- Fossil fuel burning is increasing the atmosphere's store of carbon by **6.1 Giga tones each year**

- Humans have created an imbalance in the carbon cycle significantly influencing the earths climate system

II. WHAT IS CLIMATE CHANGE?

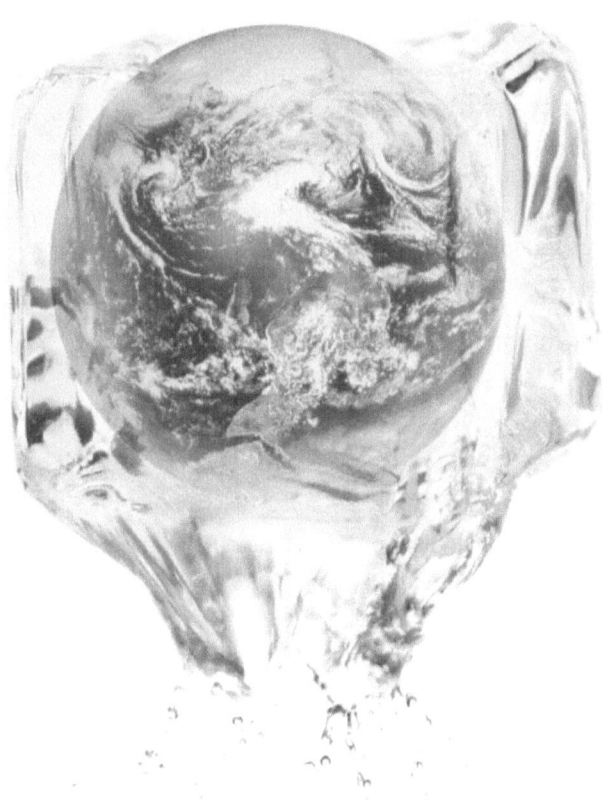

- The Intergovernmental Panel on **climate change** (IPCC) refers to climate change as a change in the state of the climate that can be identified (e.g. using statistical tests) by changes in the mean and/or the variability of its properties, and that persists for an extended period, typically decades or longer. It refers to any change in climate over time, whether due to natural variability or as a result of human activity

- The United Nations Framework Convention on climate change (UNFCCC), refers to climate change as a change of climate that is attributed directly or indirectly to human activity that alters the composition of the global atmosphere and that is in addition to natural climate variability observed over comparable time periods

- More commonly, climate change is a shift in long-term average weather patterns

- The international scientific community agrees that there has been a significant change in global climate in recent years, particularly in the polar areas

- The most noticeable and direct impact of climate change observed in recent years is **global warming**.

- Global warming is a significant increase in global ambient temperatures

- The recent warming trend is now known to be undoubtedly mainly caused by anthropic activity due to the addition of greenhouse gases into the atmosphere mostly through the burning of fossil fuels

Brief History of The Earth Climate: Natural Variability

- The Earth is a very dynamic planet which has undergone many transformations throughout its history and since its formation 4.6 billion years ago

- Climate changes have continued throughout the history of earth. A succession of geological events and the evolution of life have dramatically influenced the climate of the earth.

- The field of Paleoclimatology has provided information of climate change in the ancient past, supplementing modern observations of climate.

- Climate variability usually occurs naturally on the planet over millions of years or suddenly as a result of a catastrophic event (e.g. major volcanic eruptions, meteor impacts).

- Throughout the history of the earth, the climate has been fluctuating between cold and warm periods.

- The history of the earths' climate can be reconstituted using various indicators ranging from the fossil record and species distribution to geological formations and plate tectonics

- Main climate eras have been identified.

A Few Examples of Past Climates

Miocene Climate
- The climate during the Miocene was similar to today's climate, but warmer. Well-defined climatic belts stretched from Pole to Equator, however, there were palm trees and alligators in England and Northern Europe. Australia was less arid than it is now

Early Cretaceous Climate
- The Early Cretaceous was a mild "Ice House" world. There was snow and ice during the winter seasons, and Cool Temperate forests covered the polar regions

Early & Middle Jurassic Climate
- The Pangean continent Mega-monsoon was in full swing during the Early and Middle Jurassic. The interior of Pangea was very arid and hot. Deserts covered what is now the Amazon and Congo rainforests. China, surrounded by moisture bearing winds was lush and verdant

Early Permian Climate (280million years ago)
- Much of the Southern Hemisphere was covered by ice as glaciers pushed northward. Coal was produced in both Equatorial rainforests and in Temperate forests during the warmer "Interglacial" periods

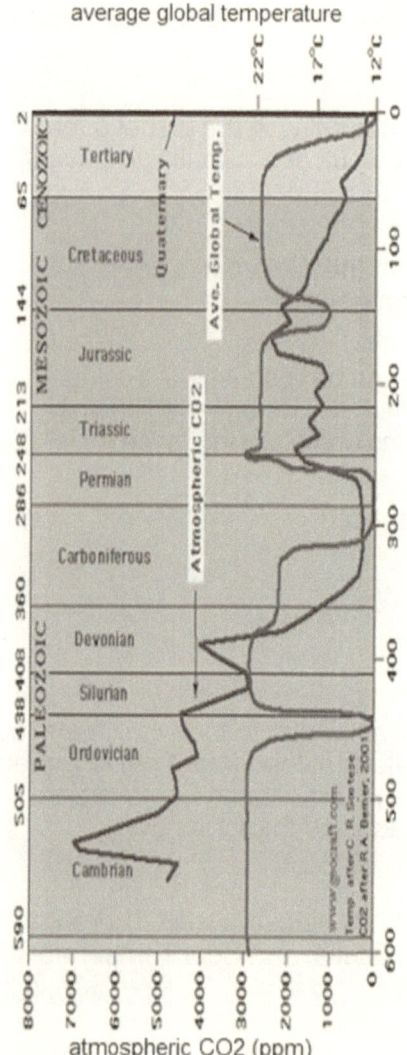

average global temperature / atmospheric CO2 (ppm)

Late Devonian Climate (360 million years ago)
- During the Late Devonian Pangea began to assemble Thick coals formed for the first time in the tropical rainforests in the Canadian Arctic and in southern China. Glaciers covered parts of the Amazon Basin, which was located close to the South Pole

Silurian Climate (420 million years ago)
- Coral reefs thrived in the clear sunny skies of the southern Arid belt which stretched across North America and northern Europe. Lingering glacial condiitons prevailed near the South Pole

Early Ordovician Climate (480 million years ago)
- Mild climates probably covered most of the globe. The continents were flooded by the oceans creating warm, broad tropical seaways

Plate Tectonics and Past Climate Changes

- Originated from two super continents (Gondwana and Laurasia) continents have since been drifting across the surface of the earth through complex geological means

- Plate tectonics played a very important role in determining the climates successions over the continents

- Due to plate tectonics, continents have changed dramatically across time and drifted across the globe (they are still drifting today)

- As continents drifted across the globe, they experienced different climate zones depending of the latitude they were located at

- Such continental climate variations driven by continental drift is a very slow process which is measured in millions of years

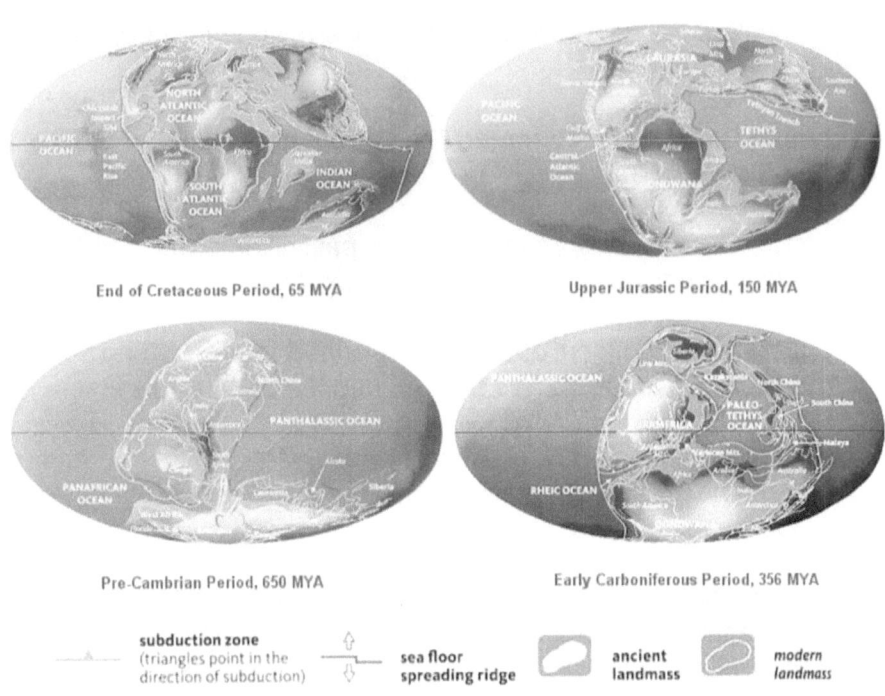

Example of A Warmer Climate: Early & Middle Jurassic Climate (206 to 144 Million Years Ago)

- The Jurassic is an example of a climate era which was much warmer than the climate that we are experiencing today.

- Much of the Jurassic world was warm and moist, with a greenhouse climate.

- Although some arid regions remained, much of the rest of Pangea was lush and green. Northern (Laurasian) and southern (Gondwanan) biotas were still distinct in many ways, but by the Jurassic, faunas had acquired a more intercontinental character. Some animals and plants were found nearly worldwide, instead of being restricted to particular regions.

- The polar areas were ice-free during this period.

- The ocean surface stood at a higher level with respect to the continents.

- Much of the earth was covered in warm and moist tropical forest

Example Of A Colder Climate: The Ordovician Climate (488 To 444 Million Years Ago)

- The Ordovician is an example of a past climate which was much colder than the climate that we are experiencing today

- The supercontinent of Gondwana drifted over the south pole, initiating a great Ice Age that gripped the earth at this time.

- Gondwana, particularly Africa, straddled the South Pole and became extensively glaciated. There were even glaciers in what is now the Sahara.

- Metazoans were severely affected. About 60% of animal genera became extinct, making this the second or third most deadly mass extinction of the Phanerozoic

Present Day Climate Change: Global Warming

- Ones more in the history of the earth, the climate is changing

- There is however several differences between past changes in the climate and what we are observing today:

Present climate change is occurring very rapidly as opposed to past climate changes which usually occurs over millions of years (rapid change have occurred in the past but as a result of catastrophic events only e.g meteorite impact, large volcanic eruptions, sudden feedback)

- Present day climate change is not a natural event but a result of human activity

- Present day climate change can be prevented as we do know the causes and how to limit the impacts

- Over the past century we have observed a change in the climate system which is unlike any past natural climate change that has been experienced over the 4.6 billion years of earth history. The difference being that we, humans, are causing this rapid shift and that for the first time in history a species has the potential to alter this change…

Present Climate Change: a Brief Historical Summary

- **1824:** French physicist Joseph Fourier is first to describe a "greenhouse effect"

- **1861:** Irish physicist John Tyndall carries out research on radiant heat and the absorption of radiation by gases and vapors including CO_2 (Carbon Dioxide) and H_2O (Water)

- **1896:** Swedish chemist Svante Arrhenius first proposes the idea of a man-made greenhouse effect

- **1938:** British engineer Guy Stewart Callendar compiles temperature statistics in a variety of regions and finds that over the previous century the mean temperature had risen remarkably

- **1955:** John Hopkins University researcher Gilbert Plass proves that increased levels of carbon dioxide could raise atmospheric temperature

- **1958:** Roger Revelle and Suess employ geochemist Charles Keeling to continuously monitor CO_2 levels in the atmosphere

- **1970:** First Earth day, US starts to produce reports highlighting concerns about global warming

- **1979:** First world climate conference held in Geneva; establishment of the World Climate Programme

- **1985:** Scientists at the World Climate Program conference at Villach in Austria confidently predict that increased CO_2 concentrations will lead to a significant rise in the mean surface temperatures of the earth. Meanwhile, a hole in the ozone layer is discovered over Antarctica

- **1987:** The hottest year on record to date

- **1988:** The Intergovernmental Panel on climate change (IPCC) is set up by the World Meteorological Organization (WMO) and by the United Nations Environment Program (UNEP). Dr James Hansen of the NASA Goddard Institute for Space Studies delivers his famous testimony to the U.S. Senate. Based on computer models and temperature measurements he is 99 percent sure that the human caused greenhouse effect has been detected and it is already changing the climate

- **1990:** The IPCC delivers its first assessment on the state of climate change, predicting an increase of 0.3 °C each decade in the 21st century -- greater than any rise seen over the previous 10,000 years

- **1992:** The United Nations Conference on Environment and Development, better known as the Earth Summit, takes place in Rio de Janeiro attended by 172 countries. It is the first unified effort to get to grips with global warming and leads to negotiations which result in the Kyoto Protocol.

- **1995:** The hottest year on record. Four years later the 1990s are confirmed as the hottest decade in 1000 years. The IPCC report for that year states that "the balance of evidence suggests a discernible human influence on global climate"

- **1997:** The Kyoto Protocol: Industrialized countries agree to cut their emissions of six key greenhouse gases by an average of 5.2 percent. Under the terms of the agreement each country -- except developing countries -- commits to a reduction by 2008 -- 2012 compared to 1990 levels

- **2001:** Newly elected U.S. President George W. Bush renounces the Kyoto Protocol stating that it will damage the U.S. economy. The third IPCC report declares that the evidence of global warming over the previous 50 years being fueled by human activities is stronger than ever

- **2003:** Europe experiences one the hottest summers on record causing widespread drought claiming the lives of over 30,000 people

- **2005:** Following ratification by Russia -- the 19th country to do so -- in November 2004, the Kyoto Protocol becomes a legally binding treaty. America and Australia continue their refusal to sign up claiming reducing emissions would damage their economies

- **2007**: 175 countries in total have ratified the Kyoto Treaty. Under new Prime Minister Kevin Rudd, Australia ratifies the treaty. The IPCC report for a fourth time states that "warming of the climate is unequivocal" and that the levels of temperature and sea rise in the 21st century will depend on the extent or limit of emissions in the coming years. Former vice-president Al Gore and the IPCC jointly win the Nobel Peace Prize for services to environmentalism.

- **2008**: 160 square miles of the Wilkins Shelf breaks away from the Antarctic coast. Scientists are concerned that Climate Change may be happening faster than previously thought. Following the Bali talks/roadmap, negotiators from 180 countries launch formal negotiations towards a new treaty to mitigate climate change at the Bangkok Climate Change Talks

- **Feb 2009**: Barack Obama becomes president of the United States. His administration is set to facilitate the instauration of Green renewable technologies, to progressively reduce dependence on fossil fuel and to strengthen international collaboration in mitigating the impacts of climate change. There is hope that the USA will join the international community in fighting climate change

- **2010:** The Cancun agreements are adopted
- **2011:** COP17 takes place in Durban
- **2012:** COP18 takes place in Doha
- **Nov 2013:** COP19 takes place in Warsaw
- **December 2014:** COP20 takes place in Lima
- **December 2015:** Paris agreement is adopted
- **October 2016:** Parliament gives its consent to the EU's ratification of the Paris agreement
- **November 2016:** Paris agreement enters into force
- **November 2016:** COP22 takes place in Marrakesh
- **June 2017:** President Donald Trump announces his intent to withdraw the United States from the Paris agreement
- **November 2017:** COP23 takes place in Bonn

...

III. CAUSES OF CLIMATE CHANGE

The Natural Causes of Past Climate Change (1): Non human Related

There are **four main** natural processes which can account for major shift in the earth climate over time:

Intense Volcanic Activity:
- Volcanoes when erupting inject a large amount of greenhouse gases into the atmosphere which can have a significant impact on the climate. The historic record shows evidence of a correlation between periods of intense volcanic activity, the level of carbon dioxide into the atmosphere and global surface temperatures.

The Milankovitch Cycles:
- They refer to the variation of the tilt of the earth's axis away from the orbital plane. The tilt varies between 22.1° and 24.5° averaging 23.5°. The obliquity changes on a cycle taking approximately 40,000 years.

- As this tilt changes, the seasons become more exaggerated. "the more tilt means more severe seasons, warmer summers and colder winters; less tilt means less severe seasons, cooler summers and milder winters."

- For an increase of 1° in obliquity, the total energy received by the summer hemisphere increases by approximately 1%

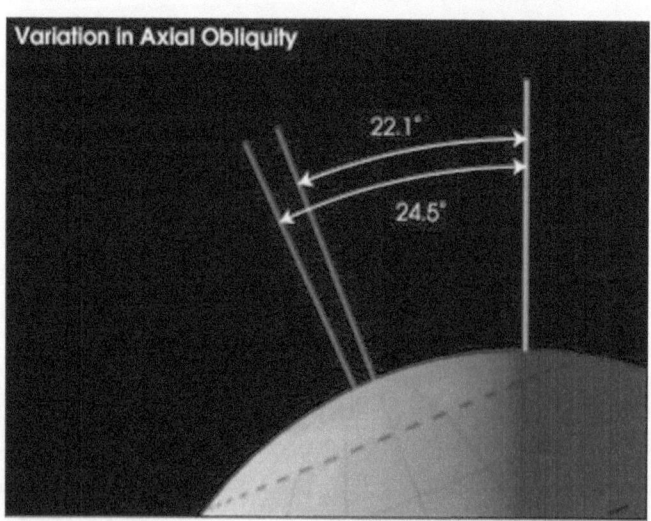

Solar activity:
- The activity of our sun is not constant

- At periodic intervals the Sun undergoes periods of intense activity which are reflected by the apparition of Solar Spots on its surface

- Intense activity of the sun has some impacts on our climate. It remains however a small influence

Tectonic Activity: Continental Drift
- The continents are constantly moving on the surface of the earth in a process referred to as plate tectonics

- In the past the continents as we know them today were very different

- Today's continents drifted away from one major continent known as The Gondwana

- While moving across different latitudes of the planet, different areas of continents passed through different climate zones

- This is however a very slow process

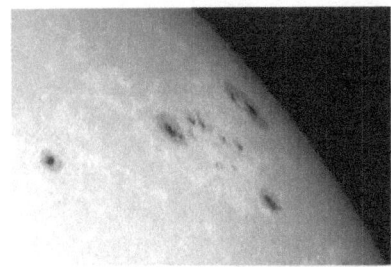

none of the natural causes combined can account alone for the rapid climate shift observed today...

Main Warming Mechanism: The Greenhouse Effect

How does it work

- Our earth receives most of its energy in the form of radiation from the sun

- The incoming solar energy (Ultra Violet UV) has a very short wavelength and passes through the atmospheric gases unaffected to reach the earth surface

- The earth surface absorbs the solar energy and releases it back to the atmosphere at a different wavelength as infrared (IR) radiation, some of which goes back into space

- Some of the IR radiation emitted by the earth is absorbed by greenhouse gases in the atmosphere.

- These gases absorb the infrared radiation emitted by the earth and re-radiate the energy as heat back towards the earth causing a warming known as the greenhouse effect.

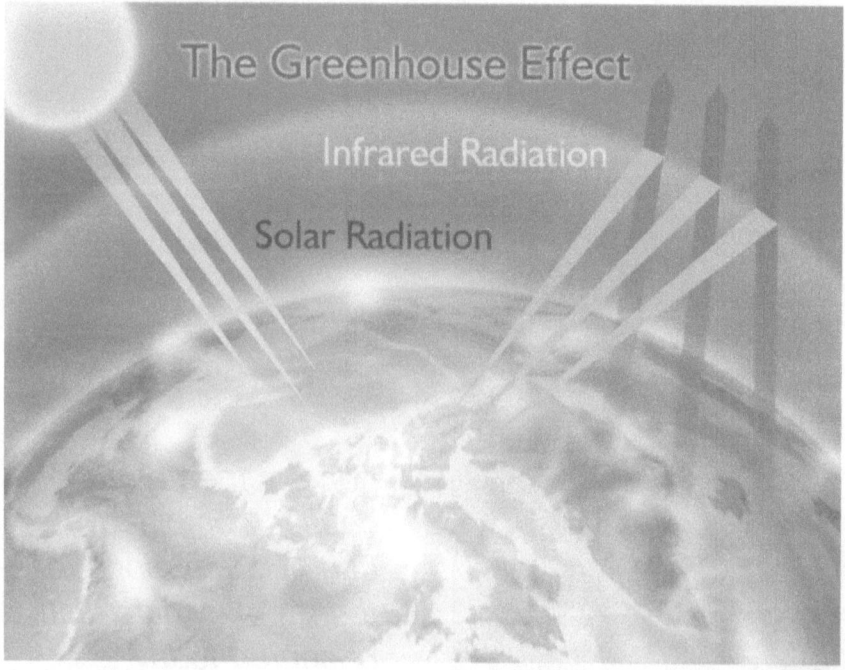

- The greenhouse effect insulates earth, resulting in the mild temperatures at the earth's surface that have allowed life to flourish. The greenhouse effect is absolutely vital to allowing life, as we know it to survive on earth. Without the greenhouse effect, earth would be a cold planet, with a mean surface temperature well below freezing.

- Thanks to the greenhouse effect the average global temperature of the earth is 15° C

- *Our current civilisation is based on a fossil fuel economy. Fossil fuel long stored in the earth crust which we now burn to generate Energy and tools (plastics). We long thought that burning fossil fuels will have no impacts on our surrounding environment. This has now been proven wrong by our understanding of the greenhouse effect. The burning of fossil fuel is indeed having devastating effects on our climate.*

Cause Of Present Warming: Principal Greenhouse Gases

Water Vapor (H_2O) : The most abundant greenhouse gas, but importantly, it acts as a feedback to the climate. Water vapor increases as the Earth's atmosphere warms, but so does the possibility of clouds and precipitation, making these some of the most important feedback mechanisms to the greenhouse effect

Carbon Dioxide (CO_2) : Carbon dioxide enters the atmosphere through the burning of fossil fuels (oil, natural gas, and coal), solid waste, trees and wood products, and also as a result of other chemical reactions (e.g., manufacture of cement). Carbon dioxide is also removed from the atmosphere (or "sequestered") when it is absorbed by plants as part of the biological carbon cycle or absorbed by the ocean which acts as a dumper

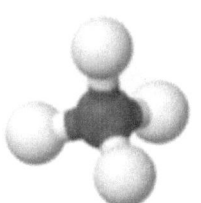

Methane (CH_4) : Methane is emitted during the production and transport of coal, natural gas, and oil. Methane emissions also result from livestock and other agricultural practices and by the decay of organic waste in municipal solid waste landfills

Nitrous Oxides (NO$_x$) : Nitrous oxide is emitted during agricultural and industrial activities, as well as during combustion of fossil fuels and solid waste

Fluorinated Gases : Hydrofluorocarbons, perfluorocarbons, and sulfur hexafluoride are synthetic, powerful greenhouse gases that are emitted from a variety of industrial processes. Fluorinated gases are sometimes used as substitutes for ozone-depleting substances (i.e., CFCs, HCFCs and halons). These gases are typically emitted in smaller quantities, but because they are potent greenhouse gases, they are sometimes referred to as High Global Warming Potential gases ("High GWP gases")

Our understanding of the physical principals behind the greenhouse effect makes it undeniable that added man made greenhouse gases are causing most of the current warming and that the temperature rise is proportional to the amount of GHG present in the atmosphere

The burning of fossil fuel, which generates the emission of greenhouse gases into the atmosphere, is making the greenhouse effect increasing stronger. One can imagine the earth as a planet surounded by a giant greenhouse (sphere of glass) and imagine what it would be like to live in such conditions. The addition of greenhouse gases in the atmosphere makes it behave more and more like a glass greenhouse. As a result of this process the earth is getting warmer and warmer

Cause of Present Warming: Greenhouse Gases

What is the warming potential of Greenhouse Gases?

- All greenhouse gases (GHG) do not have the same warming potential neither will they remain in the atmosphere for the same amount of time. This means that with the same amount of gas released into the atmosphere, some will have a much greater warming impact than other GHG

- Presently the main anthropogenic gas responsible for global warming is carbon dioxide (CO_2) due to the very large amount released. However, other gases, such as methane (CH_4), have a much greater warming potential but are currently released to a much lower extent (or in much lower concentrations)

- Scientists predict that the current warming of the atmosphere could trigger the release of large amounts of much stronger greenhouse gases, (such as methane CH_4) from land fields and ocean beds.

Gas	Atmospheric lifetime (years)	100-year GWP (Global Warming Potential)
Carbon dioxide (CO_2)	50-200	1
Methane (CH_4)	12	21
Nitrous Oxide (N_2O)	120	310
Halogenated Chloro Fluorocarbon (HFC-23)	264	11,700
Tetrafluoromethane (CF_4)	50,000	6,500
Sulfur hexafluoride (SF_6)	3,200	23,900

Source: IPCC

• For convenience, scientists use carbon dioxide as a reference and the warming potential of other greenhouse gases is expressed in terms of CO_2 equivalent (As such, CO_2 is said to have a warming potential of 1)

• For instance methane (CH_4) is 21 times more efficient at warming the atmosphere than carbon dioxide (CO_2) while Halogenated Chloro Fluorocarbon (HCF-23) is 11,700 More efficient!

• Ones released into the atmosphere, carbon dioxide will remain there for 50 to 200 years while methane will only remain about 12 years and CF_4 50,000 years!

"There is no more serious doubt that our climate is warming...
most of the warming of our climate is very likely to be due to increasing greenhouse gas (GHG) concentrations in the atmosphere resulting from human activities"

International Panel for Climate Change 2007

Cause Of Current Warming: Greenhouse Gases Sources

- Global GHG emissions due to human activities have grown since pre-industrial times, with an increase of 70% between 1970 and 2004

- The largest growth in GHG emissions between 1970 and 2004 has come from energy supply, transport and industry, while residential and commercial buildings, forestry (including deforestation) and agriculture sectors have been growing at a lower rate

- Global atmospheric concentrations of CO_2, CH_4 and N_2O have increased remarkably as a result of human activities since 1750 and now far exceed pre-industrial values determined from ice cores spanning many thousands of years

Source: IPCC

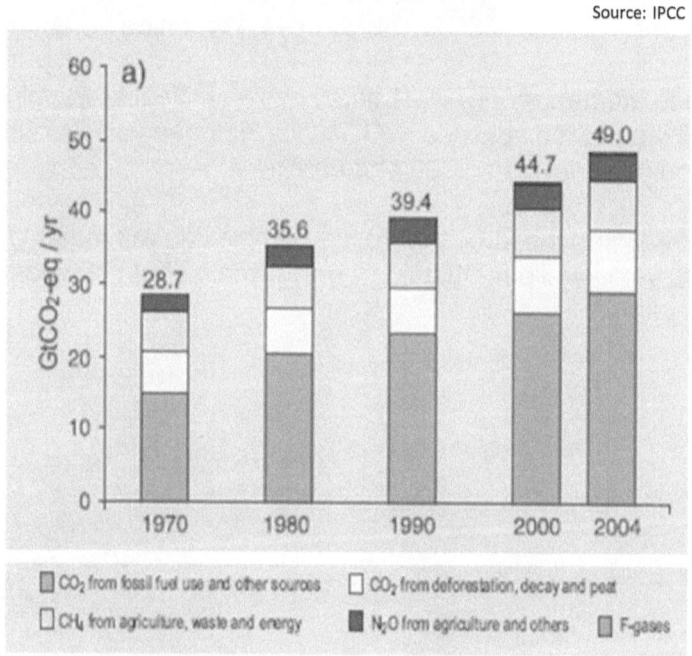

- As indicated on the left graphics, more than half of global GHG emissions come from CO_2 of which annual emissions have grown between 1970 and 2004 by about 80%

- Deforestations is the second largest source of CO_2 emissions

- Energy supply and Industry remain the 2 major sources of GHG

- Forestry, agriculture and transport are also major sources

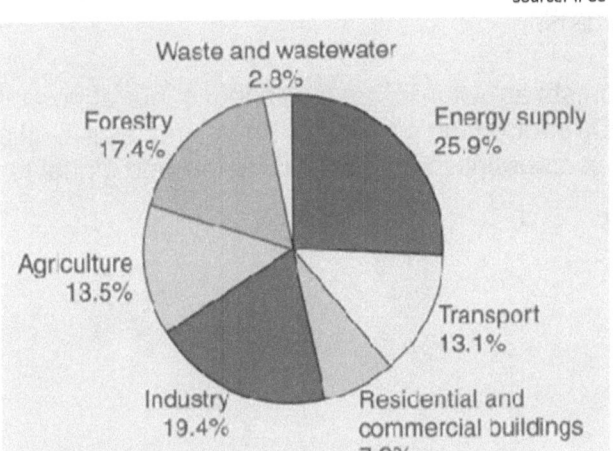

Source: IPCC

Cause Of Current Warming: Deforestation and Climate Change

- Greenhouse gas emissions from fossil fuels is by far the main cause of global warming (natural forcing included)

- However, deforestation is also an important factor which has serious implications for long term sustainability of our climate system (and biodiversity)

- Deforestation plays on two different levels:

- The loss of CO_2 uptake by forests depletion and therefore the increase in global CO_2 levels
- The decay of organic matter and the release of significant amounts of methane (a strong GHG gas) in the process

- Deforestation mostly comes from cutting down trees for commercial and heating purposes, fires, pollution (acid rains, soil and air pollution); mining; extreme events; melting of the permafrost

- Massive reforestation would be a relatively inexpensive way to limit the impacts of climate change (limit the increase of GHG and soil erosion)

- Mangroves are amongst the most efficient trees for the storage of GHG. However, mangroves around the world are rapidly vanishing

- Forests are vital to the equilibrium of our atmosphere by removing CO_2 through the process of Photosynthesis. Cutting down forests is contributing in raising GHG levels and global temperatures

IV. MAIN PROOFS OF CLIMATE CHANGE

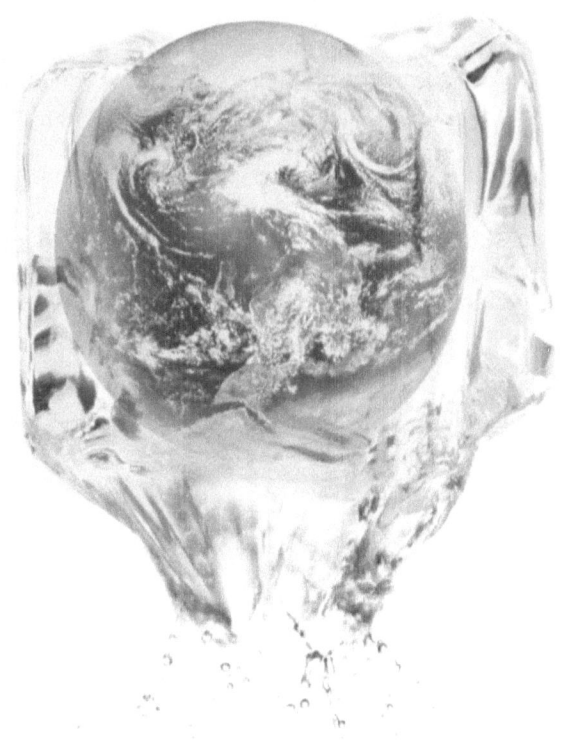

"Scientific evidence for warming of the Climate System is **unequivocal**."

(Intergovernmental Panel on Climate Change 2007)

Warnings Of A Changing Climate: A Global View

- The science of climate change has made a lot of progress over the last decades. The predictions on climate change do not come from a few individuals but from a large number of leading scientists all over the world undertaking research in different fields from geological/biological surveys to atmospheric monitoring (In 1988 the IPCC was created for this purpose)

- All the data collected from these different fields and independent studies point to the same conclusion that our climate is rapidly changing at an unprecedented rate mostly due to the action of man

- There are two levels of evidences for climate change: one which is a long term trend (the most reliable) and evidences which are already happening within our lifespan. We are starting to observe noticeable evidences and signs of climate change happening at present which are a growing concern as their frequency and strength are predicted to increase significantly throughout the century

Large blocks of ice breaking off ice shelf's and glaciers is now a common event. A clear indication of a fast changing climate. In 2008, 160 square miles of the Wilkins Shelf broke away from the Antarctic coast. Such event is expected to become more and more common

Heat Waves	In 2003 Europe experienced the hottest summers on record causing widespread drought claiming the lives of over 30,000 people. In 2009 the state of Victoria in Australia has seen its highest temperatures ever recorded (>27°C) which facilitated the spread of the worst and most deadly fires in its recent history.
Storms	More intense and frequent storms. In 2005 Hurricane Katrina swept the US Coast as a Category 5 Hurricane. 2008 was the most active hurricane season in US recorded history
Floods	In 2008 India experienced its strongest monsoon in 50 years causing widespread floods and displacing millions of people
Droughts	2008 Australia experienced its worst drought in a century which affected food prices around the world

And the list goes on.... |

Evidence : Indicators Of Past Climate, Paleoclimatology

Paleoclimatology is the study of past climate, for times prior to instrumental weather measurements. Paleoclimatologists use clues from natural "proxy" sources such as tree rings, ice cores, corals, and ocean and lake sediments to understand natural climate variability

From this proxy scientist are able to obtain precise information's on the past climate by reconstructing past temperatures and atmospheric CO_2 content on a year by year basis

- For example, ice cores removed from 2 miles deep in the Antarctic contain atmospheric samples trapped in tiny air bubbles that date as far back as 650,000 years. These samples have allowed scientists to construct a precise historical record of greenhouse gas concentration stretching back hundreds of thousands of years

- *Large blocks of ice breaking off ice shelf's and glaciers is now a common event. A clear indication of a fast changing climate. In 2008, 160 square miles of the Wilkins Shelf broke away from the Antarctic coast. Such event is expected to become more and more common*

Evidence: Indicators Of Past Climate

The CO_2-temperature correlation

- The information's collected from the Paleoclimatic studies have produced (separately) two sets of graphs. One of which represents the levels of carbon dioxide in the atmosphere and another atmospheric temperatures. The age of the samples taken can be measured accurately using isotopic dating

- These graphs clearly indicate the relationship between the level of carbon dioxide (CO_2) in the atmosphere and global temperatures are closely correlated

- This is yet again strong evidence that observed climate change is mainly caused by human activity through the addition of greenhouse gases (mostly CO_2) into the atmosphere

- CO_2 is not the only greenhouse gas (it neither has the strongest warming potential) but considering that it is largely predominant in the atmosphere comparatively with other greenhouse gases it is presently the most to blame for global warming

These two graphics show an almost perfect match between levels of carbon dioxide in the atmosphere and atmospheric temperatures. Such perfect match cannot be accidental and provides very strong evidence of a direct correlation between CO_2 levels and temperatures. Source IPCC

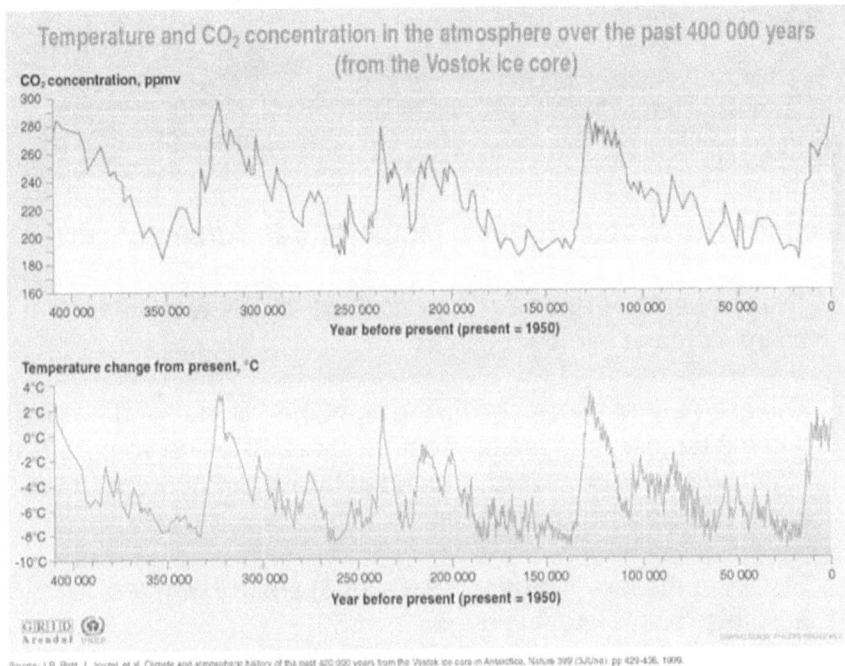

Evidence: The Atmospheric CO_2 Peak Anomaly

The comparison of Paleoclimate data and actual present days atmospheric measurements (last right section of the graphic) have identified a remarkable anomaly in the CO_2 of the atmosphere which has sharply risen since the industrial revolution

Levels of Carbon Dioxide are higher today than at anytime in the past 650,000 years

Furthermore, never in history has the level of carbon dioxide increased so rapidly

Natural causes such as volcanic activity (or solar activity) cannot explain the level of CO_2 observed today

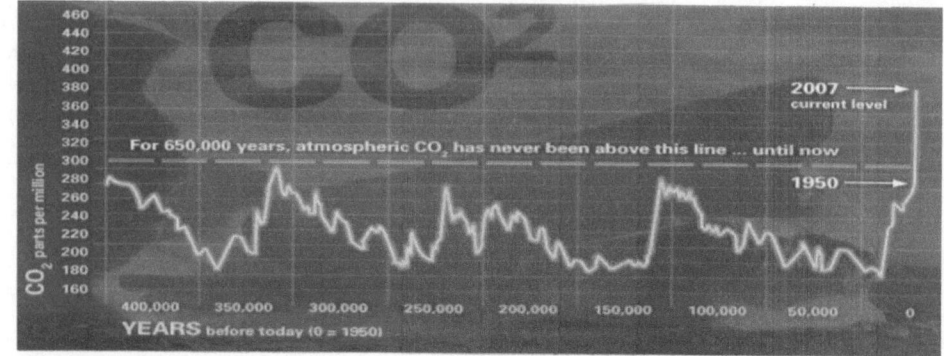

Source: NASA

Evidence: The Global Surface Temperatures Anomaly

- The temperature anomaly appears here in yellow representing a sharp increase in recent years

- The above time series shows the combined global land and marine surface temperature record from 1850 to 2007. The year 2007 was eighth warmest on record, exceeded by 1998, 2005, 2003, 2002, 2004, 2006 and 2001 in respect.

- Eleven of the last 12 years (1995-2006) are the warmest since accurate record keeping began in 1850

- Global surface air temperatures rose three-quarters of a degree Celsius in the last century, but at twice that amount in the past 50 years

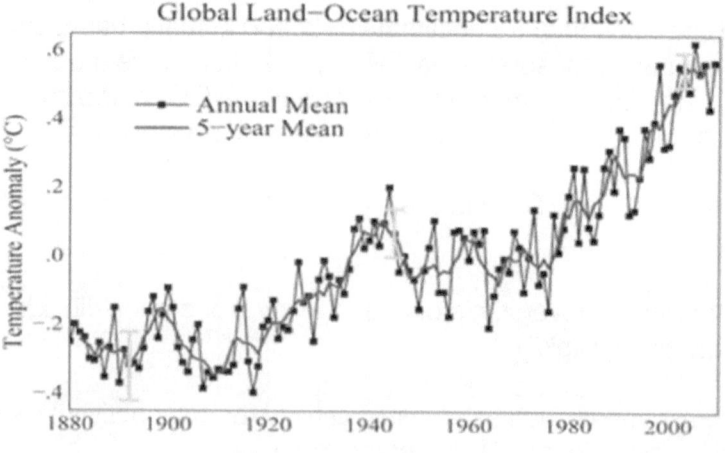

Source: NASA

Evidence: The Global Surface Temperatures Anomaly

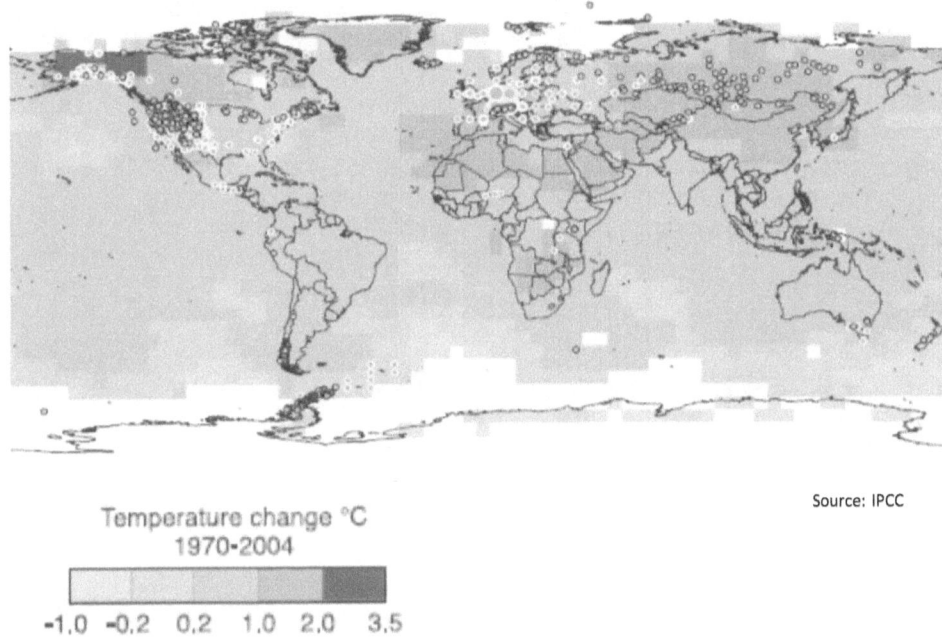

Source: IPCC

Change in Surface Temperatures (1970-2004)

- The following map indicates the extent of warming that has been taking place over the earth surface in the last 25 years

- The surface of the earth is not warming at the same rate in different parts of the world

- The high latidues have experienced the most warming and in some parts the temperature has increased by an impressive 3.5 °C

Evidence: The Computer Model

- Climate modeling which originated from meteorological modeling is a powerful tool that scientist now use to predict how our climate will evolve

- On these graphs, blue shaded bands show the 5 to 95% range for 19 simulations from 5 climate models using only the natural forcing due to solar activity & volcanoes. Red shaded bands show the 5 to 95% range for 58 simulations from 14 climate models using both natural & anthropogenic forcing

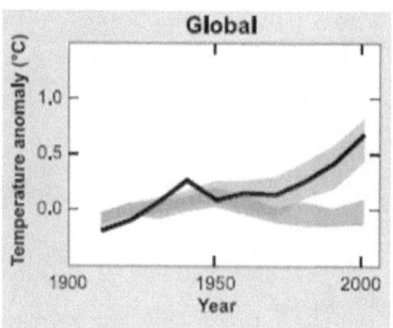
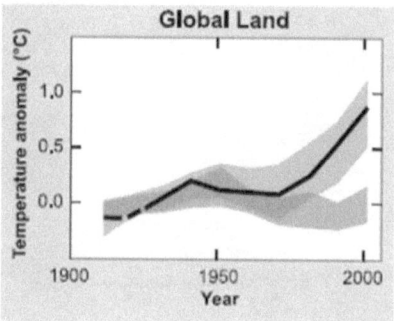

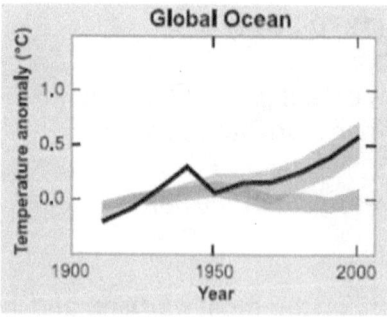
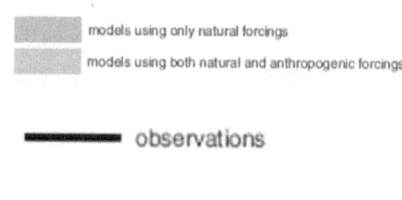

Source: IPCC

- These graphics suggest that the main natural causes of warming (solar activity & volcanoes) cannot alone explain the current trend that we observe. It also indicates that the current trend can only be accounted for when considering the human impact through the anthropic addition of greenhouse gases to the atmosphere

- This is very strong evidence that the current warming we observing is mainly caused by human activity and cannot be explained by natural forcing alone (volcanic activity and the influence of the orientation of the earth on its axis referred to as the Milankovitch cycles)

- The models well predict the observed trend when the carbon dioxide from the burning of fossil fuel is incorporated in addition to natural forcing

Evidence: Sea Level Rise

- This chart shows historical sea level data derived from 23 tide-gauge measurements. The bottom chart shows the average sea level since 1993 derived from global satellite measurements

- Sea level rise is associated with the thermal expansion of sea water due to climate warming (mostly) and widespread melting of land ice

- Global sea level rose about 17 centimeters in the last century. In the last decade, however, the rate of rise nearly doubled

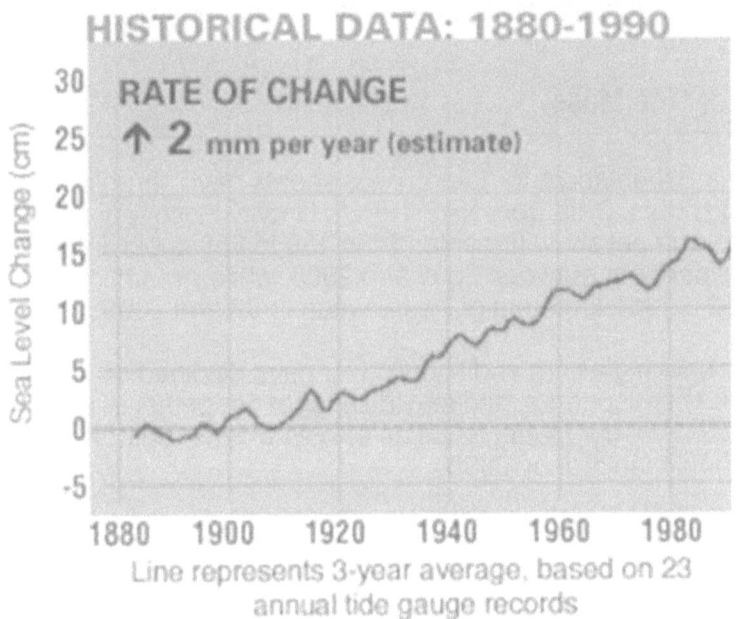

Source: IPCC

- Sea level is not rising at the same rate everywhere on the planet. This is due to the complex dynamic effects occurring on the surface of the globe (ex: sea and surface currents, local anomalies, tides, topography). There can be differences of more then a meter in different areas of the planet (ex: from one side to the other of the pacific ocean during the El Nino anomaly)

- The best current estimate is that the global sea level is presently rising at a rate of 3.4 mm a year and shows signs of increasing

- It is important to understand that these are the current values but that the rate of sea level rise will significantly increase as the global temperature rises

- The current estimates are mostly based on satellite monitoring. However, there are some great concerns from leading scientist and recent studies that these predictions are underestimating the rate of change to come as we still do not fully understand the dynamic of ice stored on the continents and how it will react to rising temperatures. It appears that ice is melting from below which cannot be picked up by satellite.

- Global sea level rose about 17 centimeters in the last century. In the last decade, however, the rate of rise nearly doubled

Evidence: Ice Mass

- The Greenland and Antarctic ice sheets have shrunk in both area and mass. Data from NASA JPL's Gravity Recovery and Climate Experiment show Greenland lost 150 to 250 cubic kilometers of ice per year between 2002 and 2006, while Antarctica lost about 152 cubic kilometers of ice between 2002 and 2005.

- Mountain glaciers and snow cover have declined on average in both hemispheres, and may disappear altogether in certain regions of our planet, such as the Himalayas, Australian Alps or the Kilimanjaro, by 2030.

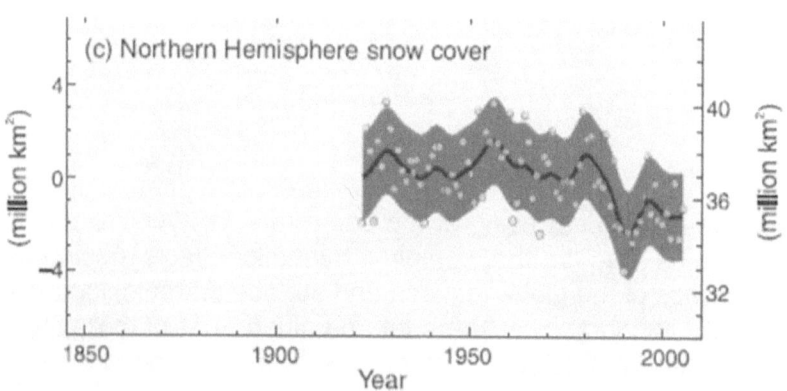

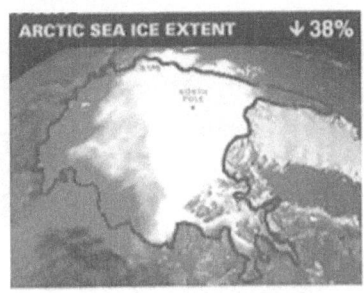 

Source NASA

2007 and 2008 were the two lowest sea ice coverage ever recorded in the arctic and recent studies warn of an increasing melting trend. The artic is very likely to be completely sea ice free within a decade (which also has important ecological and feedback consequences).

- Sea ice melt is a clear evidence that the sea and atmospheric temperature are warming
- When sea ice melts it does not contribute to global sea level rise, however land ice directly contributes to a rise in sea level when melting.
- There are 5-6 meters worth of sea level in the Greenland ice sheet, and 6-7 meters in the West Antarctic Ice Sheet which is of great concern if there were to melt (which has already began)

Evidence (8): Ocean Salinity Variations

- Precipitation and evaporation patterns over the oceans have changed, increasing ocean salinity near the equator and decreased salinity at higher latitudes
- As more water evaporates near the equator, the upper part of the oceans in these areas increase in salinity as a result
- On the other hand, as its ice melts, the excess fresh water dilutes the salinity in higher latitudes (which could have major consequences for the oceanic circulation patterns)
- Salinity variations provide additional evidence of a warming Planet

Evidence (9): Atmospheric Water Vapor

- As Global atmospheric temperatures rise so do sea surface temperatures
- The rise in sea surface temperature results in a more intense rate of of evaporation resulting in atmospheric water vapor levels to rise
- The increase in Atmospheric water vapor, is another evidence of rapid warming of our planet

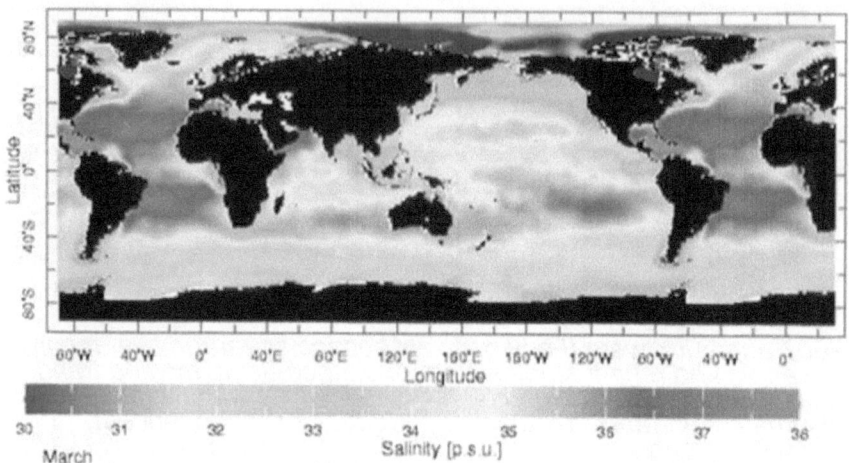

Surface Salinity Patterns Source: Lamont Doherty Earth Observatory, Climate Modeling and Diagnostics Group

ATMOSPHERIC WATER VAPOUR

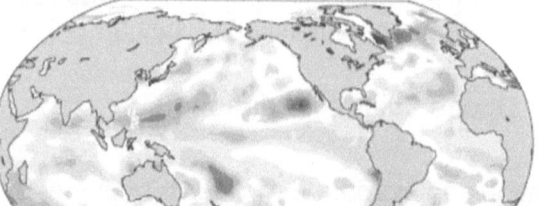

Source: IPCC

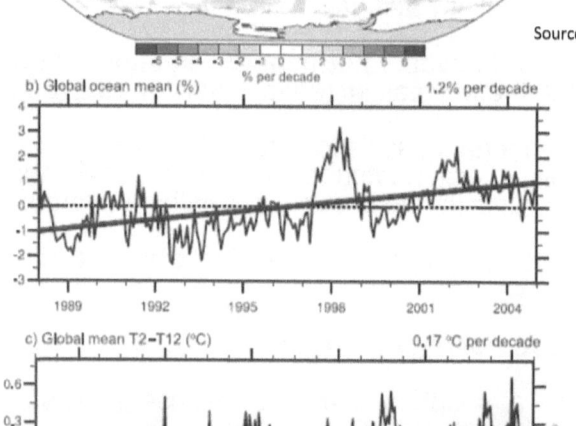

Various Other Indicators Of Climate Change

In addition to these major indicators providing strong evidence that our Climate is rapidly changing, several indicators have been developed and can serve as proxies for further evidence:

Indicators	Tendency observed
Date of Leaf Emergence on Trees in Spring	Appears sooner
Dates of Insect Appearance and Activity	Appears sooner
Abundance of plankton	More abundant
Upstream Migration of Salmon	sooner
Egg-laying Dates of Birds	Sooner in the season
Vineyards - areas under production	increased
Seasonal Patterns of Human Mortality	More apparent and increasing
Number of Outdoor Fires	Increases
Soil conditions	Dryer in some parts and more humid in other parts of the world

And many more...combining all the data from these different observations, there is no doubt that our planet is rapidly warming

V. PREDICTING FUTURE CLIMATE THROUGH MODELING

Modeling The Climate

- Climate modeling derives from the science of meteorology

- The atmosphere and upper part of the oceans are sudivided into cells

- Each cell is considered individually using a complex branch of mathematics known as integral calculus

- Very powerfull computers running millions of caculations per second enable us to model the natural processes influencing our climate and weather

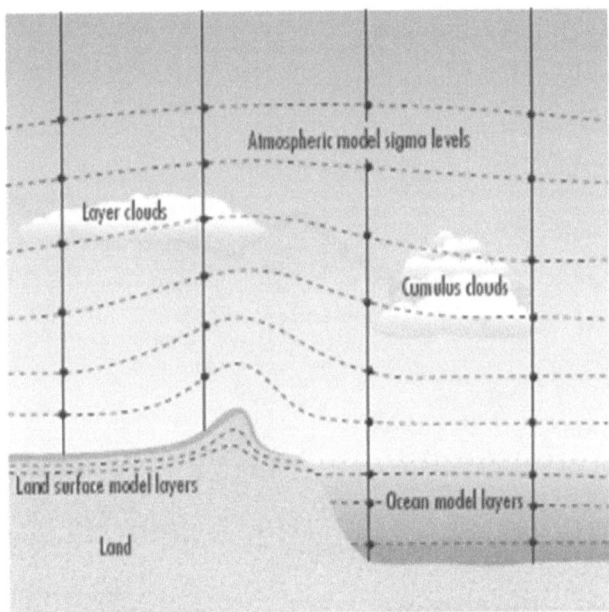

Accuracy of Computer Models

Computer models are widely used to predict the evolution of the climate system but how predictive are they?

Blue shaded bands show the 5 to 95% range for 19 simulations from 5 climate models using only the natural forcing due to solar activity & volcanoes. Red shaded bands show the 5 to 95% range for 58 simulations from 14 climate models using both natural & anthropogenic forcing

- This graphic suggest that the computer models used by scientists to predict the climate are accurate and match the actual observations that we have observed over the past century

- It also suggest that since the models have been so accurate at predicting what is actually happening they will continue to predict with a good level of accuracy what will happen to our climate over the century under different emission scenarios

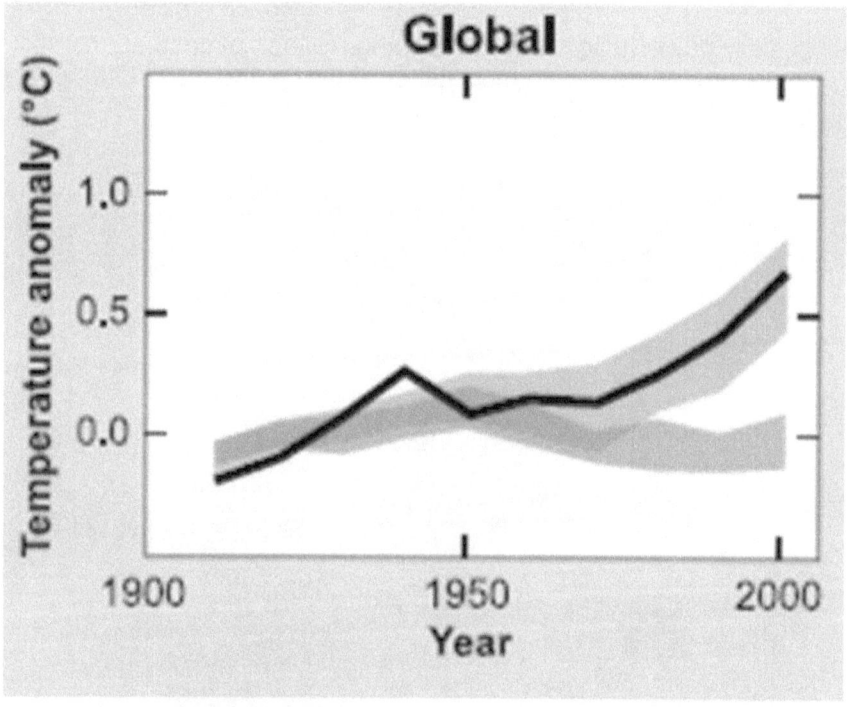

Source: IPCC

GHG Emission Predictions

Scenarios for GHG emissions from 2000 to 2100 in the absence of additional climate policies

- This graph shows six (6) illustrative Special Report on Emissions Scenarios (SRES) and an 80th percentile range of recent scenarios. The emissions include CO_2, CH_4, N_2O and Fluorinated gases

- All models predict a steady increase in GHG emissions until mid century
- Some models predict that GHG will sharply decrease after mid century mostly due a depleting oil reserves
- Some models predict that oil reserves will be used to the limit and continue to increase until the end of the century
- Some models based on recent discoveries predict that other GHG will take over as a result of feedback effects from rising temperatures

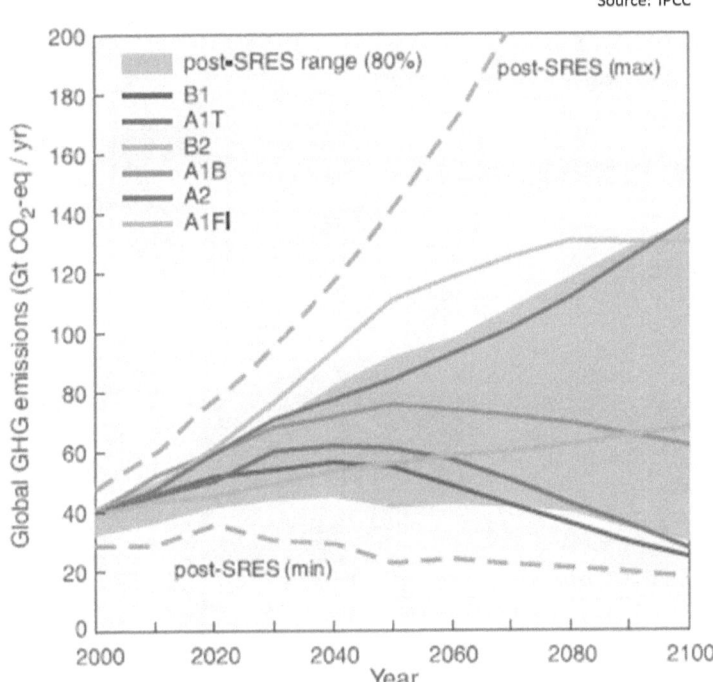

Source: IPCC

Sea Level Rise Predictions

All current models agree on a rise of about half a meter by the end of the century

- Since most of the global population lives on costal areas, sea level rise will have a major effect on world population displacing millions of people

- Sea level rise will also have drastic impacts on water resources (salt contamination of fresh water supplies)

- Agricultural lands will become infertile and flooded

Projected global average sea level rise at the end of the 21st century

Case	Sea Level rise (in m at 2090-2099 relative to 1980-1999)
B1 scenario	0.18-0.38
A1T scenario	0.20-0.45
B2 scenario	0.20-0.43
A1B scenario	0.21-0.48
A2 scenario	0.23-0.51
A1F1 scenario	0.26-0.59

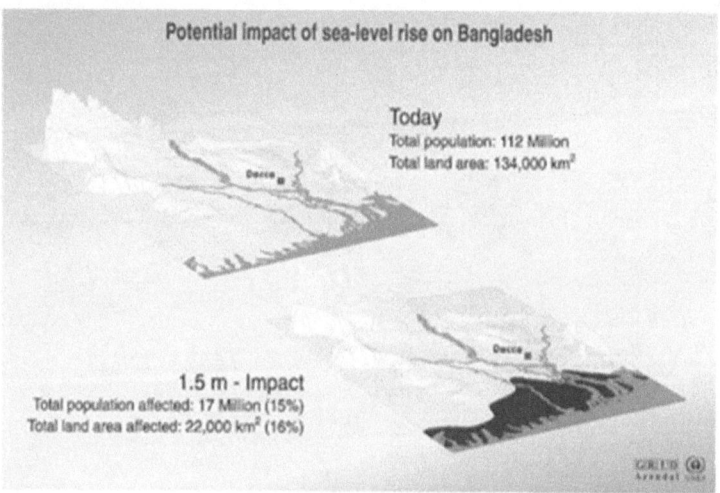

This illustration shows a modeling of what will happen to the coast line of Bangladesh under a 1.5m rise (source IPCC)

Case	Atmospheric CO_2 equivalent (ppm)
B1 scenario	600
A1T scenario	700
B2 scenario	800
A1B scenario	850
A2 scenario	1250
A1F1 scenario	1550

Source: IPCC

As an indication, the different scenarios represent the above atmospheric GHG concentrations by the end of the century

Surface Temperatures Predictions

For the next two decades a warming of about **0.2 °C per decade** is projected for a range of SRES emissions scenarios. Even if the concentrations of all GHGs and aerosols had been kept constant at year 2000 levels, a further warming of about 0.1 °C per decade would be expected. Afterwards, temperature projections increasingly depend on specific emissions scenarios

Warming by 2090-2099 relative to 1980-1999 for non-mitigation scenarios

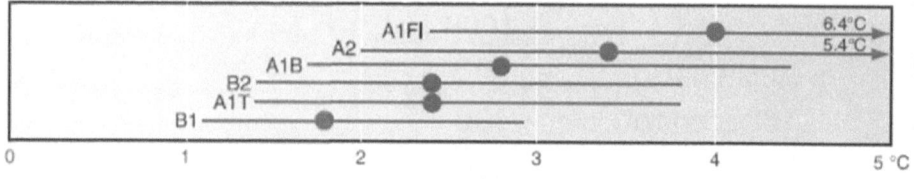

Temperature Change (°C at 2090-2099 relative to 1980-1999)

Case	Best estimate	Likely range
Constant year 2000 concentrations	0.6	0.3-0.9
B1 scenario	1.8	1.1-2.9
A1T scenario	2.4	1.4-3.8
B2 scenario	2.4	1.4-3.8
A1B scenario	2.8	1.7-4.4
A2 scenario	3.4	2.0-5.4
A1FI scenario	4.0	2.4-6.4

Source IPCC

Precipitations Predictions

•There is an improving understanding of projected patterns of precipitation.

• The impacts that climate change will have on precipitations patterns is uneven. Some areas will experience more rain while others will experience much less leading to severe droughts.

• Increases in the amount of precipitation is very likely in high-latitudes, while decreases are likely in most subtropical land regions (by as much as about 20 % in the A1B scenario 2100) continuing observed patterns in recent trends

• Globally the change in rain patterns could lead to severe potable water shortage especially in poorer countries

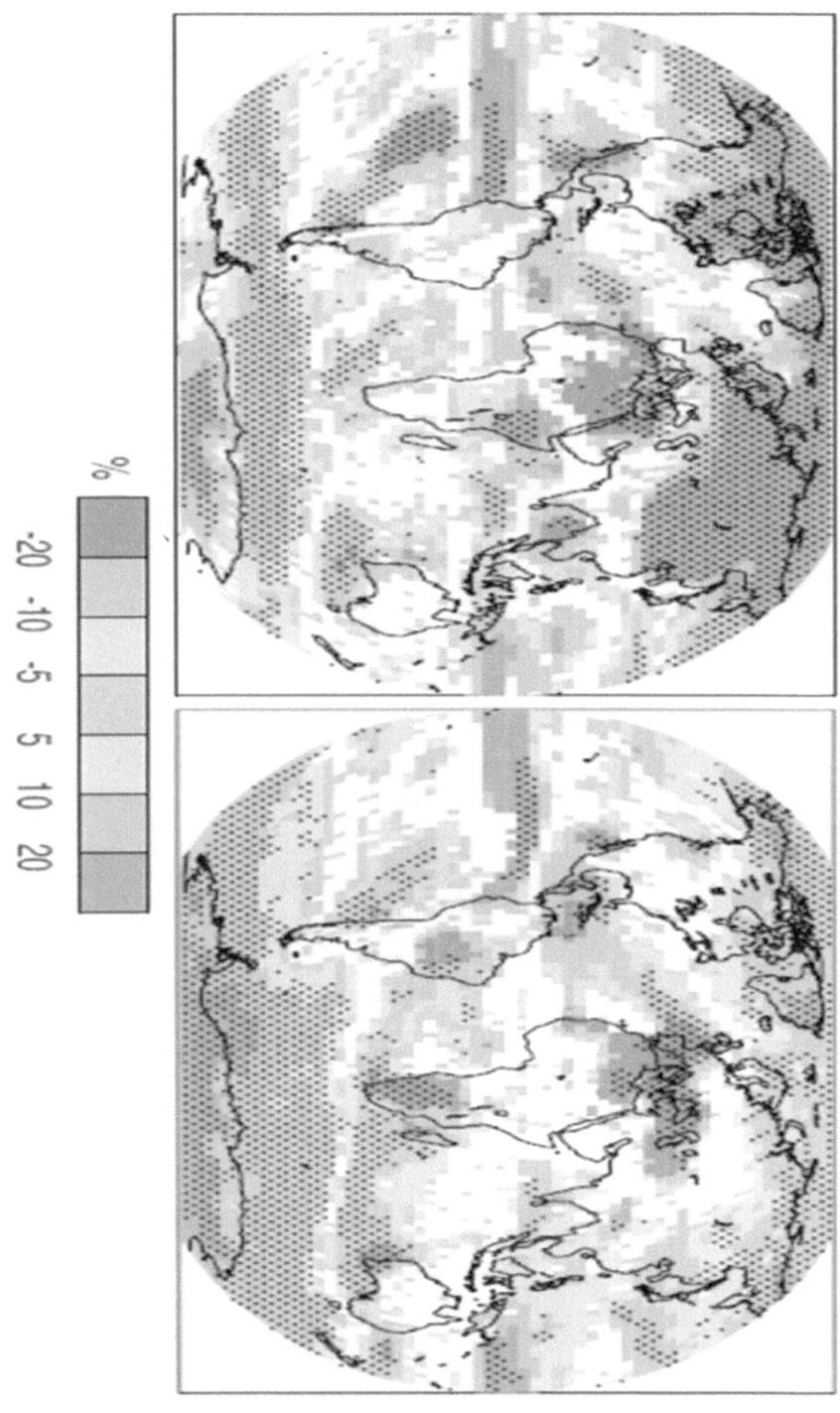

Source: IPCC

VI. IMPACTS OF CLIMATE CHANGE

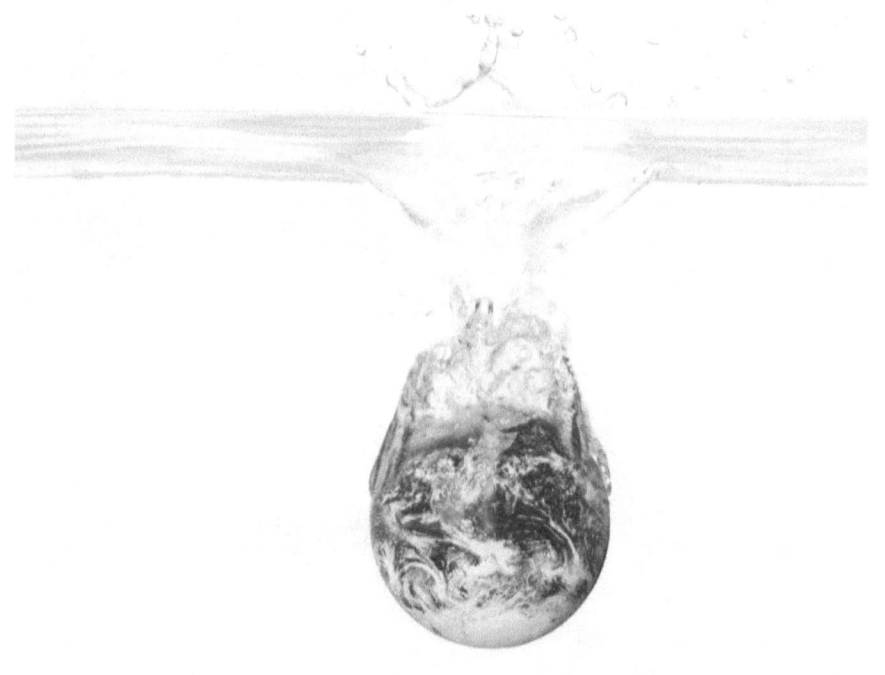

Impacts of Climate Change on Biodiversity

- We now know from our understanding of biology, ecology and other life sciences that many organisms on our planet are very sensitive to variations in temperature and will not be able to adapt to the extremely rapid changes which usually occurs over millions of years

- Many terrestrial animals are already under threat solely on the basis of temperature rise

- Approximately **20 to 30%** of plant and animal species **assessed so far** are likely to be at increased risk of extinction if increases in global average temperature exceed 1.5 to 2.5°C

- According to a recent study by the IUCN (October 2008) **35%** of the worlds birds, **52 %** of amphibians, **71 %** of warm-water reef building corals are likely to be particularly susceptible to climate change

- Global warming also induces species migrations. Many species of plants and animals are already responding to global warming, moving to higher elevations or closer to the poles

- Many scientist believe that we are on the verge of a massive species extinction event which would be catastrophic and have serious consequences for us human beings. Mass extinctions of this magnitude have only occurred five times in the history of our planet; the last brought the end of the dinosaur age.

- Species in the high latitudes are particularly at early risk (loss of biodiversity will not be limited to high latitudes) as the weather in the poles (especially the North Pole) is warming up three times faster than the rest of the planet

- Species like seals and polar bears become victims of disappearing ice shelf's as they now have to swim greater distances off the coast to find food supplies

- The loss of biodiversity will be one of the most dramatic consequences of climate change. Unlike human societies which can adapt to the changes (in some parts), numerous species will simply disappear from the earth. One must remember that extinction of species is forever and cannot be reversed

- Biodiversity is a great value to our planet and should be preserved at all costs. Loosing biodiversity will not only mean a loss of beauty and diversity but also substantial economical loss in terms of commerce and potential medicines and knowledge

- Species are already disappearing at a much greater rate then they are studied. While extinction is a natural process, human impacts have elevated the rate of extinction by at least a thousand, possibly several thousand, times the natural rate

- We live in a complex, interlocking, symbiotic relationship with other organisms, and our ecosystem is a fragile one in which our own survival depends on those of other species

- Arguments such as economic development must prevail over the preservation of species is no longer acceptable if we are to sustain a long term healthy environment and ecosystems. There is nothing less durable and sustainable that the definitive disappearance of species from our planet. Acting to preserve them is a moral and ethical obligation for future generations.

- We could loose 40% of the worlds total biodiversity from the impacts of climate change alone by the end of the century!

Mangroves are vulnerable to sea level and temperature rise and are likely to disappear along with their very rich biodiversity

- Many species have a very restricted repartition range (referred to as micro endemism) around micro habitats such as a single mountain or a lake. The disruption of these habitats as a consequence of climate change will also mean the disappearance of their biodiversity

- Climate change will result in significant loss of habitats

Biodiversity Hotspots:

In a world where conservation budgets are insufficient given the number of species threatened with extinction, identifying conservation priorities is crucial. British ecologist Norman Myers defined the biodiversity hotspot concept in 1988 to address the dilemma that conservationists face: what areas are the most immediately important for conserving biodiversity:

Marine Ecosystems

- Marine environments will be greatly affected by climate change

- As the atmospheric temperature rise ocean surfaces temperatures will also rise

- A slight rise in ocean temperature coupled with acidification from CO_2 uptake will cause entire coastal ecosystems to change and possibly collapse

- Large migration of species will occur

- Many species unable to migrate to warmer areas fast enough will become extinct

- In 2008 massive blooming of jellyfish where recorded across the world. In some parts the blooming was found to be due to warmer surface temperatures as a result of global warming.

- Such events will become more frequent and will have serious consequences on marine ecosystems such as a decline in fish populations

Impacts of Climate Change on The Oceans

- The oceans are a great sink of atmospheric Carbon Dioxide CO_2 by absorbing a large amount of it

- However, the increased concentration of dissolved carbon Dioxide in the oceans makes them become more and more acidic. About a third of man-made carbon dioxide emissions has dissolved into the oceans. As carbon dioxide dissolves in seawater, it forms carbonic acid, which lowers the ocean's alkalinity and pH level, making it more acidic

- Even a slight acidification of oceans will have an impact on biodiversity and will put lots of species at risk

- Chicago University researchers recently found that acidity levels increased at more than 10 times the rate predicted by computer models designed to study the link between atmospheric concentrations of carbon dioxide and ocean acidity

- Acidification is also likely to impair the movement and function of high oxygen demand fauna (e.g. squid, fish)

- Acidification of the oceans will have a great impact on the ability of organisms to produce calcium carbonate necessary for their survival

Coral bleaching is an early sign of the impacts that climate change is having on ocean biodiversity. The death of corals could trigger entire reef ecosystems to collapse and many other species to become extinct

- A large proportion of our food supplies comes from the oceans. Ocean acidification has the potential to substantially destabilize all marine ecosystems which will add onto already extreme pressures put on those ecosystems from over fishing, overexploitation of resources and pollution

Impacts of Climate Change on Water Resources

- Another major concern of climate change will be increasing pressures on fresh water supplies

- Climate change will significantly affect the distribution of rain around the globe

- While some areas will drastically start running out of water others will experience frequent flooding due to stronger monsoons and heavy rainfalls (both of these events lead to a diminution of available water)

- Groundwater storage is already running out in many parts of the world. Many of these supplies are also contaminated and unusable

- Some areas will become much dryer, raising concerns about the availability of fresh water for millions of people (e.g. Central Africa)

- Many lakes will dry out as a result of increased temperatures and evaporation

Depletion of potable water resources could become a major security threat and humanitarian crisis especially in poor countries which cannot afford alternative technologies

Impacts of Climate Change on Societies

- More than 80% of the worlds population will live on the coast lines and in urban areas by 2050

- Sea level rise/floods and droughts will displace millions of people creating environmental refugees and is likely to trigger world instabilities leading to conflicts for resources

- Many agricultural lands will be flooded and even greater surfaces are likely to become contaminated by salt making them unusable for agriculture

- Some small island states with low elevations (< 2m) will entirely disappear (ex: The Maldives)

- Some low lying countries such as Bangladesh and the Nederland's will loose very significant landmass

- In addition to loosing significant land mass, many of the most threatened places are also in areas vulnerable to tropical storms. The storm surges during such events will have even more devastating effects.

- Some cities like Venice and New Orleans are already experiencing serious troubles due to sea level rise and are responding by spending billions on preventive measures (but what will happen to cities and countries which cannot afford such measures?)

Small Island States

Small Island nations will be amongst the first affected by climate change

They will suffer from:

- Major surface area loss

- Coastal erosion

- Serious depletion of water supplies

- Rise in infectious diseases

- The numerous low lying island sates will generate environmental refugees from different nationalities.

Where will these refugees go?

Adaptability will become a key factor if we are to succeed in the fight against climate change. Societies will have to adapt to a rapidly changing environment

Impacts of Climate Change on Human Health

- The health status of millions of people is projected to be affected through, for example, increases in malnutrition; increased deaths, diseases and injury due to extreme weather events; increased burden of diarrhoeal diseases; increased frequency of cardio-respiratory diseases due to higher concentrations of ground-level ozone in urban areas related to climate change

- Increase in global temperatures will significantly increase the occurrence and severity of smog events in urban areas

- This will significantly reduce air quality

- Babies and the elderly will be most affected by a decrease in air quality

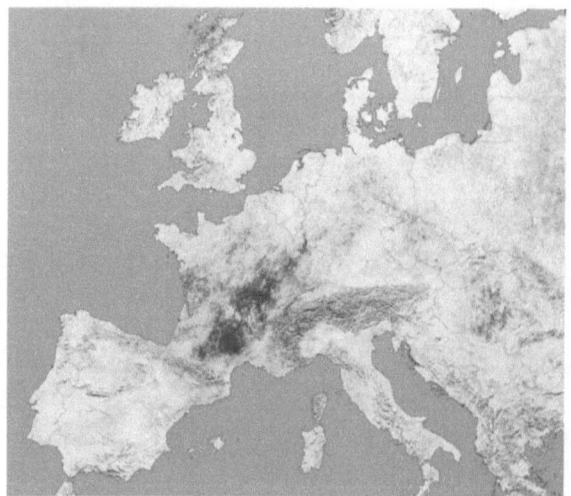

European heat wave of 2003 which killed hundred of people. Such heat waves are more likely to kill a larger amount of people in cold countries which are not used to hot weather than in hotter countries which have the infrastructure to deal with over heat throughout the year.

- Change in climatic and weather patterns will affect the distribution of infectious diseases

- Many viruses have restricted areas of occurrence due to natural physical barriers which are governed by temperature gradients

- Diseases such as Malaria and Dengue will move north as temperature rises

- The number of occurrence of infectious diseases will significantly increase

- A rise in the case of people infected by such diseases has already begun to occur in some areas which were not of concern a few decades ago

- As indicated on the graph, the number of case of reported Dengue is sharply increasing worldwide. Many scientist believe this is a direct consequence of rising temperatures

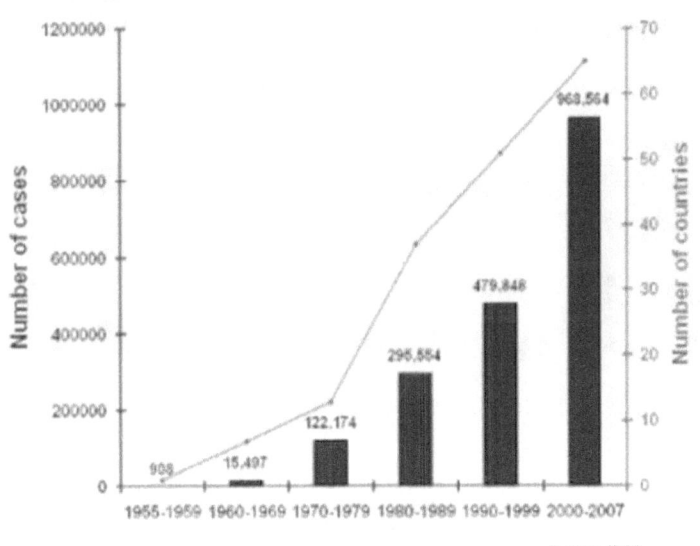

Source: IPCC

- The spread and increase of vector born diseases such as Dengue and Malaria are becoming a serious health threat as a result of increasing temperatures and change in weather patterns...

Increase of Extreme Weather Events

- It is likely that tropical storms (cyclones) will become more frequent and more intense as a result of increased sea surface temperatures and moisture in the air

- Sea level rise will have devastating effects during storms due to surges

- Based on a range of models, it is likely that future tropical cyclones (typhoons and hurricanes) will become more intense, with larger peak wind speeds and more heavy precipitation associated with ongoing increases of tropical sea-surface temperatures.

- There is less confidence in projections of a global decrease in numbers of tropical cyclones. The apparent increase in the proportion of very intense storms since 1970 in some regions is much larger than simulated by current models for that period.

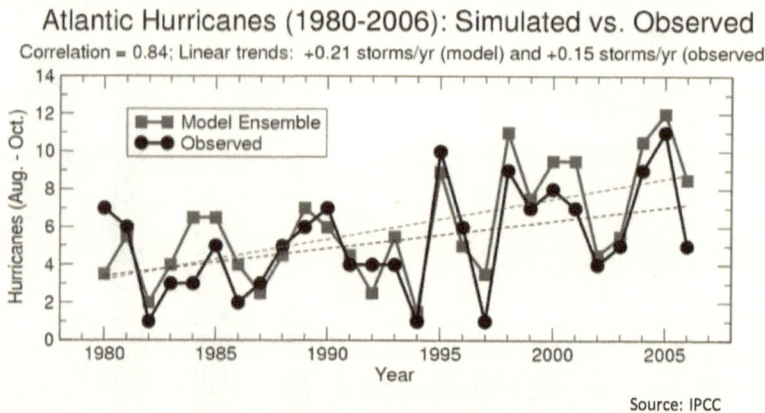

Source: IPCC

*The **2008 Atlantic hurricane season** was the **most active in recorded history**. Three records showed the hurricane season's relentlessness. Six consecutive named storms struck the U.S. mainland, something that had not been seen in recorded history. It's also the first time a major hurricane, those with winds of at least 111 mph, formed in five consecutive months, July through November and Bertha spun about for 17 days, making it the longest lived storm in July*

- Extra-tropical storm tracks are projected to move poleward, with consequent changes in wind, precipitation and temperature patterns, continuing the broad pattern of observed trends over the last half century

- Some recent studies suggest that erratic weather events have serious impacts on wildlife and will further contribute to a decline of biodiversity

- More frequent and extreme weather events as a result of rising temperatures could displace millions and push nations economies to their limit

- Estimated global macro-economic costs in 2030 and 2050. Costs are relative to the baseline for least-cost trajectories towards different long-term stabilisation levels *Source IPCC 2007*

Stabilisation levels (ppm CO_2-eq)	Median GDP reduction[a] (%)		Range of GDP reduction[b] (%)		Reduction of average annual GDP growth rates (percentage points)[a,a]	
	2030	2050	2030	2050	2030	2050
445 – 535[d]	Not available		<3	<5.5	< 0.12	< 0.12
535 – 590	0.6	1.3	0.2 to 2.5	slightly negative to 4	< 0.1	< 0.1
590 – 710	0.2	0.5	-0.6 to 1.2	-1 to 2	< 0.06	< 0.05

Economic Effects Of Global Warming

- While fossil fuel resources (petroleum) are running out (within 20 years), the price of energy is expected to rise significantly, which is likely to trigger another economical crisis such as the 2008 global economy crisis. Unless governments rapidly find solutions and plan for a progressive shift to sustainable energy use, economic instabilities appear inevitable in the decades to come.

- All economical models point to the same conclusion: the cost of doing nothing to cut GHG emissions and limit the impacts of climate change will far exceed the cost of mitigation and long term planning.

- Global warming has been identified as a major threat to world economies. If natural disasters become stronger and more frequent, nations economies will be pushed to the limits, creating great instabilities – recovery efforts for natural disasters such as Huricane Katrina cost billions of dollars. Lets just imagine what the cost would be for such events occurring on a regular basis and in different locations.

Great natural catastrophes 1950 – 2005
Overall losses and insured losses

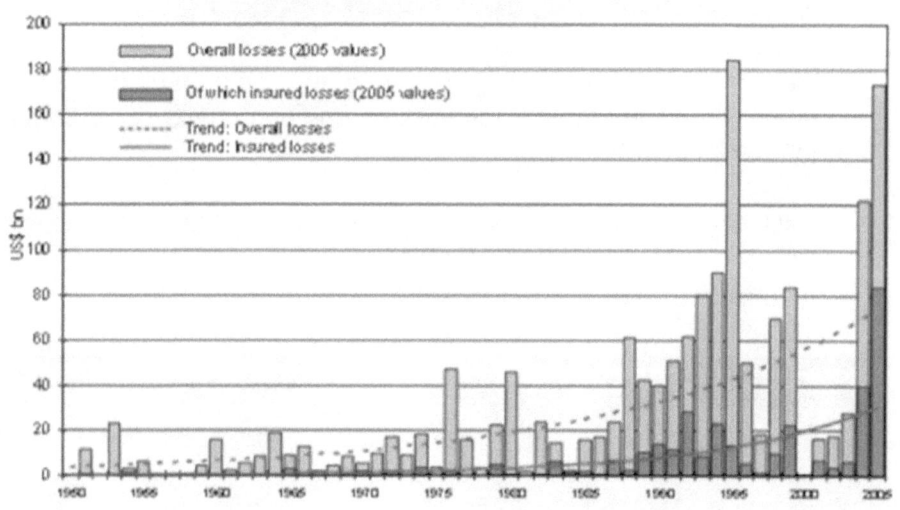

Source: IPCC

- As an example, the insurance industry lost 10.6 billion US dollars due to the 2008 Hurricane season alone

- Some very detailed studies have been published providing all the tools and guidance needed by governments to take necessary actions while there is still time to significantly reduce the impacts (e.g. the Stern report)

- We know that as the climate continues to get warmer, the number of major catastrophes will significantly increase

- We also know from past experience that the economic losses resulting from such events will be tremendous for nations economies

- According to this data, if the number of extreme events increases to the level predicted, nation economies will be devastated

- Mitigation and preventive measures would help to limit such economical impacts

Case Study: The Economic Implications of Sea Level Rise

Since the industrial revolution, megaprojects have been largely synonymous with mining, transportation (e.g. tunnels, bridges, roads) and energy infrastructures. However, this is about to shift to an era where the next generation of Large Complex Projects will be environmental-related. This area is also where future (or imminent) significant funding and business opportunities will be generated.

What is a LCP?

While a multitude of construction projects are taking place around the world at any given time, a handful of them is large and complex. Typically, large complex projects would be defined as operating on very significant budgets (not uncommonly in the USD billion-dollar range), large in scale, technologically and logistically challenging. Currently, projects falling under this category may include large bridges, tunnels, energy power plants, petrochemical infrastructures, highways, railways, mining sites, piping installations, roads or, as a matter of fact, any other projects where complexity stems from their very large scale.

Why environmental megaprojects are set to increase significantly

It is a fact that environmental issues are on the rise and the consequences of inaction are becoming more and more apparent. While climate change negotiations have and still are lagging on global actions to take, the level of warnings is unequivocal. Increased frequency of heatwaves, major storms, major drought...The signs could not be clearer and well in line with predictive climate models made over 20 years ago.

This increase in natural disasters which can only get worse as we progress throughout the century will also drive unprecedented demand for mega engineering projects, on a scale that has never been seen before, to cope with serious environmental risks.

The Panama Canal, the English Channel Tunnel or the Five Gorges Dam, all which have their names on the hall of fame of megaprojects, will seem small in comparison to the scale of engineering mega projects that will be required to cope with global change throughout this century. If we are going to have to protect island nations or entire stretches of coastlines we are talking about projects that will be visible from space on the scale of the Great Wall of China!

Type of Environmental megaprojects that will be required

While the type of projects required will need to be adapted depending on the locations, there are obvious trends that we no longer can afford to ignore:

Megatrend 1: protective infrastructures for sea level rising. This element will be the main challenge facing many cities throughout the world. The threat originates on two fronts: first, as the climate continues to warm thermal expansion and ice melt will induce significant sea-level rise. Secondly, the global population is rapidly becoming predominantly urbanized. Both of these will drive a serious demand for large scale coastal cities protective actions. When we talk about sea-level rise preventive infrastructures it can relate to a diversified range of solutions including giant sea walls, raising coastline highs through land reclamation, some dams and canal technologies, giant pumping stations...

Loss in property value due to sea level rise and flooding

Projected losses by zip code, 2005 to 2033

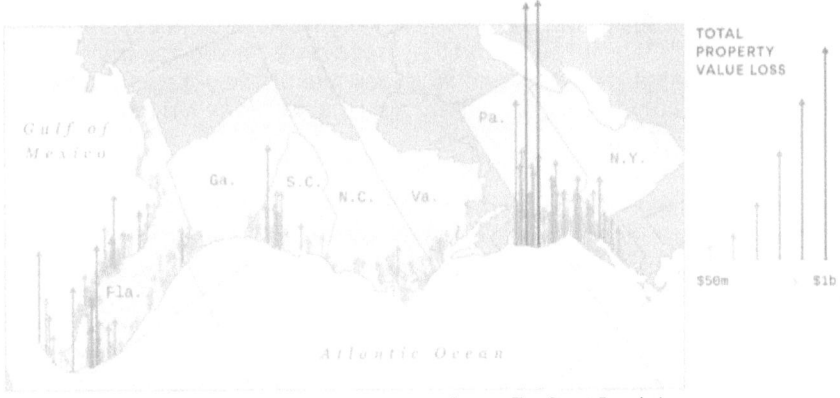

Source: First Street Foundation

Megatrend 2: Renewable energy mega factories. While the era of fossil fuels is coming to an end, the demand for alternative energies of various types is set to increase significantly in a race for a low carbon economy. These projects will include gigantic solar farms, onshore and offshore wind farms and other types of lower carbon energy coming into the mix. It is expected that different countries will adopt different solutions which are most tailored to their needs and resources. Those projects often differ from current powerplant projects by their wider footprint and the need to integrate them better into the landscape

Megatrend 3: subterranean infrastructures. It is expected that significant subterranean infrastructures will be required ranging from giant canals to drain water on a massive scale to the need for new urban energy efficient infrastructures. A good example is the flood prevention subterranean infrastructure of Kuala Lumpur which is already in existence. Improvements in tunnelling technologies learnt from current tunnel mega projects will be required.

Megatrend 4: water, waste technologies and infrastructure. Access to freshwater, treatment of waste-water and adequate management and recycling of all sorts of waste will become a significant issue and protecting existing resources and accessing new ones through technology will require large scale and ingenious technology.

A common characteristic of those projects is that they do impact the community more than conventional megaprojects and therefore, stakeholder involvement and political considerations are more important.

A trend for national defensive actions rather than global collaboration

While a clear failure can be observed to make the necessary changes on a global level through international political agreements, some governments are already taking actions into their own hands. For instance just last month the Prime Minister of the Republic of Singapore announced during his annual address to the nation that the cost of mitigating the impacts of climate change on the island from aspects such as sea-level rise would require the imminent set-up of an SGD 100 billion fund and that mega projects such as the constructions of dams, sea walls and land reclamation should start today and continue through a long term deployment plan that will span the next 50 years with annual reviews.

Certain countries like Singapore are now putting climate change on the same level as national security as highlighted by the fact that Singapore's National Climate Change Committee falls under the Prime Minister's office, and this is a very significant mindset shift in itself! The latest is not the result of some sudden environment realization but strong policy resulting from years of impact research and a lot of government money spent. In other words, Singapore politicians seem to agree that action on climate change-related impacts to come are an absolute survival necessity. The Singapore case study should both serve as an example and a warning to the world. Countries do not plan this kind of funding and strategic development without strong reasons...

Similarly, the city of Jakarta which is sinking has already established plans to build a giant sea wall megaproject as with Venice in Italy. Concurrently, Monaco is starting the construction of giant floating suburbs to extend its limited land but also to deal with rising sea levels. There are many other similar serious environmental engineering mitigation projects which are either already occurring or in the definition stage and this is just a beginning.

Preventing the worst effects of climate change, while still possible, through international policy in cutting down Greenhouse Gas emissions will no longer prevent very significant and impactful changes from occurring throughout the century. The only option is now adaptation which needs to start now, and this will only be possible through LCPs tailored to environmental engineering. But how much is all this going to cost and who is going to fund these megaprojects?

A possible funding mechanism

As the cost of financing such projects will be tremendous and take a heavy toll on the global economy, it is expected that at some point this century most LCPs will be environment-related should it be in energy infrastructures (renewables), coastal engineering or others.

Funding this type of projects in the 100s of billion range over a relatively short period will be very challenging for global and national economies.

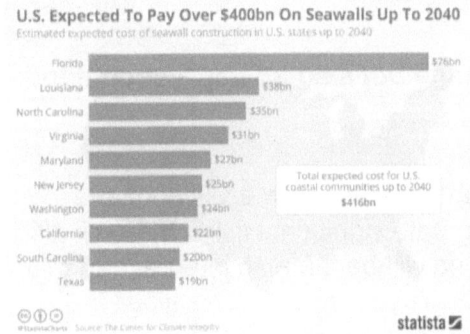

The strategies of trying to convince stakeholders to invest for future generations to prevent the impacts to come does not work well enough. Despite actions to date from a few development banks, the private sector and financial institutions still predominantly only provides funding for short terms and high return on interest projects.

Perhaps the solution is to find a way to finance projects for the long term while also benefiting current needs and generating profits. Below are some examples of this short-long term approach applied to LCPs:

Case study 1: Singapore has developed a unique dam to close the Marina Bay: The Marina Bay Barrage. This project has several benefits that include transforming the original bay seawater into a large additional freshwater reservoir and protecting the coastline against sea-level rise. While clearly planned for the long term (protecting the cost lines and providing additional water security), the project was also developed with many short terms benefits in mind which includes: a water history museum located within the dam regularly visited, playgrounds for family getaway, a range of facilities for food and beverages and shops. In other words, this long-term development has been integrated with facilities where activities prosper and overtime help to pay back the cost of the development.

Case study 2: *Cleveland City Planning Division* has developed ecological corridors to enable the migration of wildlife. While this project in itself would have been difficult to fund, the developer has incorporated a range of community infrastructures such as cycling lanes, children playground and sports facilities. The latest has driven a lot of interest and funding for the project considering that these infrastructures also benefit the needs of the current communities.

To fund projects of this scale it is primordial that short term goals be incorporated within the longer-term agenda.

Conclusion: A significant challenge but also an unprecedented opportunity

In the end, while global environmental deteriorations will continue to occur throughout the century and create significant risks, they will also create opportunities which in the end will become an integral part of the solution.

The LCPs that, for now, are mostly tailored at improving our way of life will shift at some point to become predominantly protective measures against mounting environmental risks.

The cost of financing these projects will be tremendous but if we start to incorporate a short-long term approach to financing these mega infrastructures this would give us a significant head start.

Case Study: Financing Climate Change

Securing funding for sustainable development goals is often a challenge considering the fact that such projects need to adopt a long term vision and do not always directly impact human's daily lives in a very obvious way. Perhaps a hybrid approach of combining long term sustainable development goals with other measures that have direct short term benefits on current society would have greater chance of success.

One of the challenges of financing sustainable development projects is the long term perspective of such initiatives. Indeed, sustainable development looks into views in the medium to long term when in fact most modern society models are based on short term returns or benefits. This simple and fundamental mismatch is slowing down sustainable development goals significantly.

Sustainable development is defined as development that meets the needs of the present (short term) without compromising the ability of future generations (long term) to meet their own needs. We have lost track of this short-long term relationship in sustainable development strategies.

There are many examples we could use to illustrate the problem. For instance, if we analyse the predictions of global sea level rise in response to global warming trends, many drastic actions requiring significant investments should be taken today to avoid large-scale social risks in the future. These projects of gigantic proportions would involve aspects such as re-engineering coastlines, changing construction codes or even progressively relocating entire sensitive coastal communities. However, none of these are presently taken very seriously by governments (as they require a long term approach) and we continue to build on the coast lines or reclaimed land (short term approach) regardless of the fact that there is a high probability that these investments are doomed in the medium term, based on sea level change projections alone.

It all comes down to the reality that political agendas or business investment most often have very short term goals which are dominated by duration of mandates or the fastest financial return model.

The need for long term sustainable development goals

It has been very clear for a long time that the lack of actions to prevent issues such as global climate change, poverty, biodiversity losses or various types of chemical pollution will significantly impact current, and to even greater extents, generations to come. However, many of the worsening impacts (e.g. sea level rise) are either very progressive or will only start to become significantly apparent into the future.

While adaptation strategies now seem inevitable, prevention should still be on the agenda (to lessen the severity of the impacts to come) and new strategies to finance much needed long term development plans need to be applied. The hybrid short-long term approach is perhaps the best option and the less damaging to the economy.

The need for short term financially viable goals

While long term goals are needed, the point here is that it is becoming increasingly difficult to fund such projects if there is no direct financial returns and benefits to current communities.

It is apparent that one of the failures of sustainable development strategies has been to separate the two concepts of short and long term strategies when in fact the solution relies in a continuity and synergy between the two.

While it is difficult to obtain funding for sustainable development plans, short term high-added value projects that benefit current communities and can contribute to the economy are abundant and finding a way to tap on these projects to meet longer term plans is key if viable sustainable development strategies are to be achieved.

The approach of 'investing for the future' separately of current needs is clearly one that hasn't been successful. It is clear that arguments such as 'we must invest in generations to come rather than the current needs' or 'we won't see the impacts now but it is for the greater good of future generations' despite true and much needed are outdated and unsuccessful. What we need is an approach that clearly benefits the present while at the same time having a longer term agenda and this should be a systematic approach to development.

There are two human psychological factors that development goals should pay much more attention to: first, we don't react well to long term progressive threats (e.g. climate change) and second, we are much more effective at short term goals than long term ones.

Identifying opportunities

There are many opportunities to link short term benefits with long term goals.

Sports is one of the greatest and easiest targets for obvious reasons: it is popular, it directly benefits the people (health and entertainment benefits) and it drives funding.

A good case study to illustrate the short-long term approach is the community cycling lanes combined with green corridors.

There are several successful examples of cycling lanes which have been developed around an ecological conscious concept. These cycling-pedestrian developments are surrounded by green corridors that actually have a significant benefit in terms of allowing species to migrate between parks. It is unlikely that these green corridors would have taken place (or, in other words, obtained the required funding) without the human benefit factor provided by the sport facilities. Short term goal: providing sport infrastructures for people to exercise. Longer term goal: insuring a sustainable management of biodiversity resources by providing habitats and connectivity for species to migrate between these habitats.

Tourism is another priority target sector. Indeed it is an interesting one because it is easy to use the short-long term model through the development of eco-tourism. Eco-tourism is a fast growing trend with high financial potential but at the same time provides means for government and the private sector to invest in environmental conservation.

A final word

We must find ways to better utilise mainstream development goals with direct short term impact on current human societies to also serve sustainable development goals as they are more likely to be successful. The link between short and long term goals must become much more comprehensive and apparent in developments strategic planning. We shouldn't keep on making distinctions between short term financially viable projects and longer term sustainable development objectives but on the contrary systematically incorporate longer term objectives within short term ones.

For sustainable development strategies to spread, a longer term agenda should always be hidden behind short term goals.

VII. UNCERTAINTIES AND ONGOING RESEARCH ON CLIMATE CHANGE

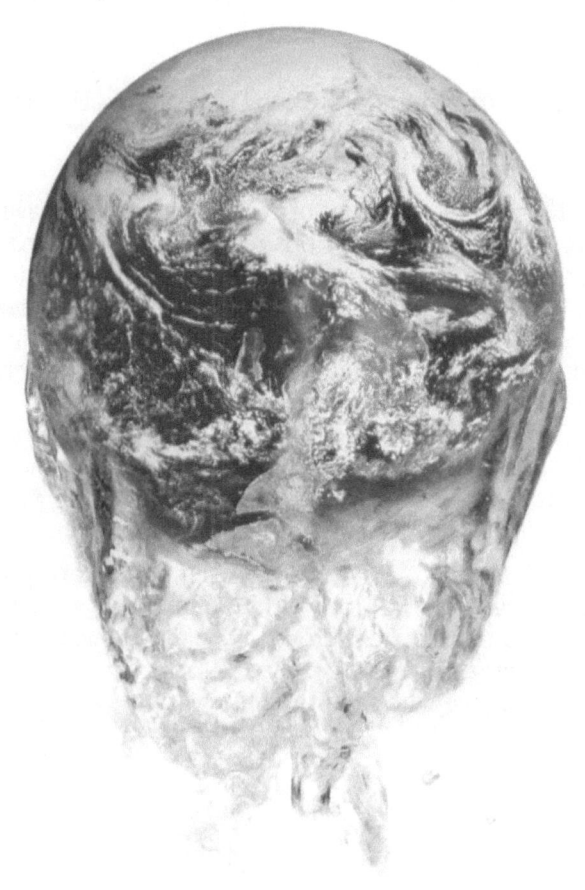

Feedback Effects

A feedback effect is a 3 step process closed loop system in which a modification of one process triggers another mechanism, which results in the amplification or reduction of the initial process.

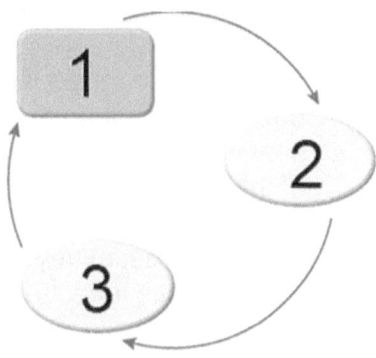

What are feedback effects?

- The climate system is very complex. Even though we can now predict with a good level of confidence the overall trend and mechanism of the climate system, there are still elements which are unpredictable with accuracy.

- Not only do increased concentrations of greenhouse gases affect the atmosphere, but also the oceans, soil and biosphere. These effects are still not completely understood.

- Feedback that results in an increase of atmospheric temperature are referred to as **positive feedback** whereas those that actually cool down the atmosphere are **negative feedbacks**. In some cases positive and negative feedbacks offset each other, which results in a state of equilibrium. However, scientists are becoming more and more concerned that in a number of systems, positive feedbacks will take over which will result in a significant increase of the current warming trend.

- Most systems are now in net positive feedback, creating an interactive set of mutually-reinforcing subsystems. This 'second-order' feedback system therefore accelerates climate change. Existing models, which do not take into account feedback effects, are therefore likely to be an underestimation of the problem.

Feedback Effects : The Albedo Effect

- Snow and ice reflect incoming solar radiation. If there is less ice due to melting then the exposed surfaces (land mass or the open sea) will absorb more energy which will result in a greater warming of the planet.

- Ice is the most extreme case of the Albedo effect, as its bright white surface reflects light (radiation) very efficiently.

- However, the same Albedo principle also applies to any surface on the planet.

- In general, the brighter the surface, the more efficient radiation will be reflected. The darker the surface, the more heat is absorbed.

Feedback Effects : Methane

Permafrost

Another feedback is the melting of permafrost in boreal regions, resulting in the release of methane, a potent greenhouse gas, and CO_2 from soil organic matter. Recent studies in Siberia, North America and elsewhere have documented the melting of permafrost.

Ecosystems

An important feedback is the release of carbon from ecosystems due to changing climatic conditions. The modification of high-carbon ecosystems, such as the Amazon, due to changes in regional precipitation patterns, has been predicted from some models, but it has not yet been observed. Laboratory studies have indicated accelerated decomposition of soil organic matter in temperate forests and grasslands due to temperature and precipitation changes, or the CO_2-induced enhancement of decomposition by mycorrhizae.

Feedback Effects: The Double Feedback Potential of Clouds

Example 2: The cloud cover

• Clouds are made of water vapour, a greenhouse gas

• The warmer the surface of the planet, the more water vapor (therefore clouds) in the atmosphere

• Water vapor is a tricky GHG as it has a simultaneous effect of warming and cooling (It reflects incoming solar radiations during the day (albedo effect), but also acts as a Green House Gas by re-emitting incoming IR rays from the Earth's surface.

• Clouds polluted by aerosols from industrial emissions have more numerous and smaller drops, causing the cloud to become brighter and reflect more of the sun's radiative energy

•The effect that clouds will have on future climate change is very difficult to predict

• Clouds are one of the strongest yet less predictable feedback mechanisms

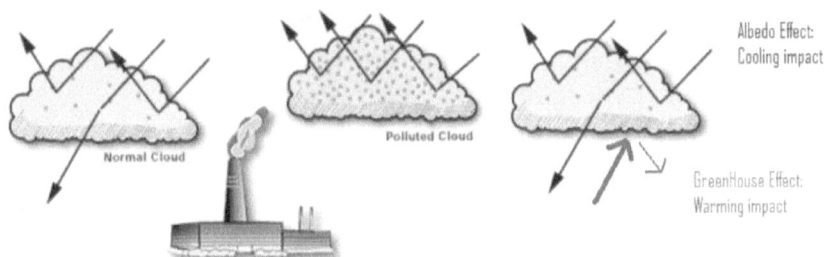

•The double cooling and warming property of clouds and its variability makes them one of the most difficult feedback mechanisms to predict

Case Study: The Threat of Feedback Mechanisms

While we tend to plan for the future based on current climate inputs and observations, we should also look ahead and take into consideration the dramatic turn of events that could result from positive feedback mechanisms.

A brief overview of feedback mechanisms

A good simple example of feedback mechanism is the steam engine: pressure builds-up inside the boiler as a result of steam generated from the heat; when the pressure passes a certain threshold, a pressure valve releases some steam in order to prevent the chamber from exploding. It is a constant adjustment process juggling between pressure build-up and pressure release; if the pressure does not build-up enough, the engine does not work but if the pressure builds up too much the system collapses....

Feedback mechanisms are processes which are the direct consequences of other events. Basically, there are two types of feedbacks: positive feedbacks which amplify the on-going trends and negative feedbacks which soften them. As an illustration, two examples of simple feedback mechanisms are given below:

<u>Positive feedback:</u> The warmer the climate (e.g. heat waves), the more energy we consume (e.g. air conditioning), the more GHGs are emitted and therefore the warmer the climate will get. We tend to forget that human behaviour in response to a changing world is in fact a powerful feedback mechanism.

<u>Negative feedback:</u> The warmer the climate; the more water vapour in the atmosphere at any given time; the more clouds which have white surfaces reflective of incoming solar radiations; the cooler the temperatures. Note however that water vapour is also a strong greenhouse gas and therefore also has a positive feedback effect.

Natural equilibrium versus destabilisation

Over billions of years of Earth history, the planet has been subject to a series of dramatic events. In between such events, systems on Earth have reached equilibriums which are governed by closed loop systems and a balance between positive and negative feedback mechanisms.

The current trend however is for human activities to modify some of these natural equilibriums by increasing disorder in the Earth systems.

Many signs of upcoming changes from positive feedbacks are already highly noticeable and measurable (e.g. release of methane gas from the melting permafrost; sea ice melt exposing darker ocean surfaces in the Arctic…). All seem to indicate that positive feedback mechanisms are on the rise and will intensify throughout the century, largely overtaking negative feedback effects which would normally act as a buffer in the systems.

The uncertainty

Most climate models (which shape policy decisions) are based on predictions which account for a range of inputs such as emissions scenarios (global amount of GHGs emitted); however few if any seriously consider how systems will evolve when feedback mechanisms are added into the equation.

The main reason is that systems are incredibly complex. Modelling some of these impacts is proving very difficult, especially certain aspects such as atmospheric water vapour that have both positive and negative feedback effects.

However, reasoning on logical deductions and how systems are likely to react should lead the way to anticipation and a more cautious approach. Some of these positive feedbacks really have the potential to shift the situation from currently alarming to dramatic and irreversible in a short time.

An unwanted but possibly unavoidable solution

If positive feedbacks become a factor as predicted by many climate scientists, our only option may end up being deliberate human negative feedback on a massive scale. We may be forced to attempt to counter warming trends through technological inputs, the most extreme of these actions being Earth Engineering. Earth Engineering, which is the process of humans voluntarily modifying natural systems on a global scale, is however not without serious consequences and is often considered a last resort.

It may seem like science-fiction but ideas such as injecting sulphur compounds into the upper atmosphere; producing artificial fine clouds on a large scale; fertilising oceans to absorb carbon dioxide; deploying a reflective membrane in the Artic or even sending millions of small mirrors into space to divert the sun rays are ideas which have been proposed and seriously considered to cool down the Earth.

The business case

However, in addition of being extremely risky, all of these engineering measures are also astronomically expensive and far more costly than mitigation measures such as cutting down current GHG emissions and shifting to a low carbon economy.
It is clear from an economical point of view that the cost of dealing with the consequences of climate change will far exceed the cost of mitigating it. This argument has been given over and over again by some leading economists (e.g. The Stern report).

Furthermore we should act fast because we are running out of time. Indeed, because GHGs remain in the atmosphere for extended periods of time, they have been accumulating. We are reaching the point of no return after which the current focus on reducing GHG emissions will no longer make much difference on the outcome (i.e. even the smallest effects will already be dramatic).

Once this point is passed, the next step will probably be to try to remove certain GHGs from the atmosphere on a large scale (which may simply take too long) followed by the final option of earth engineering.

The more we wait, the more extreme and outlandishly expensive the solutions become.

These actions are the ingredients we need to fuel a new economic revolution based on a low carbon economy. The threat of positive feedbacks should be seen as yet another argument to renew our efforts.

Uncertainty on The Melting of The Permafrost and Ice

- Some recent studies suggest that large amounts of methane hydrates stored in the permafrost will be released in the atmosphere as it melts, which will accelerate the rate of global warming.

- The melting of the permafrost will also add a significant amount of water to the oceans, which will contribute to global sea level rise.

- As the permafrost melts, soils become waterlogged and unstable. Entire towns built on these type of soils are at risk. There are also major structures worth billions of dollars installed on the permafrost, such as pipelines and electrical lines.

- Recent studies link increased pollution levels with ice melt. Soot or black carbon darkens the ice and makes it soak up more heat, accelerating melting rates compared to reflective snow and ice. Methane comes from sources including oil and gas and agriculture while ozone is formed from industrial pollutants

Ice Dynamics

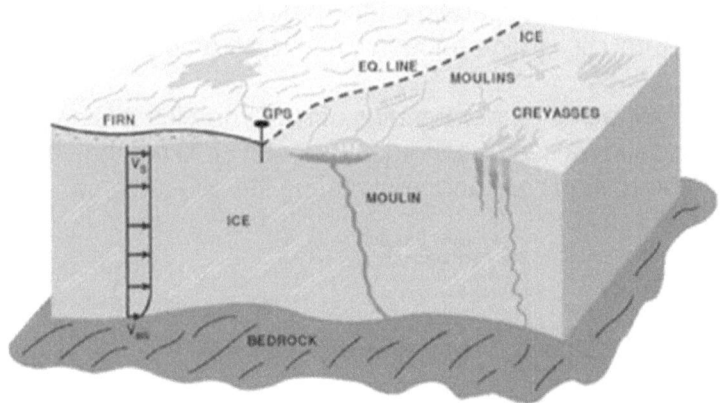

Ice melting in the arctic. These ice cascades are not ordinary and very concerning. What this picture shows is actually ice melting and going down into cavities straight to the bottom of the ice shelf. This suggest that these huge volumes of ice are melting within which is not currently incorporated into the models for sea level rise.

- The predictions regarding sea level rise over the next century remain uncertain to some extent and are likely to be underestimated.

- In its 2007 Fourth Assessment report, the Intergovernmental Panel on Climate Change used new satellite data to conclude that shrinkage of ice sheets may contribute more to sea level rise than it had thought as recently as 2001. The panel concluded that it could not "provide a best estimate or an upper bound for sea level rise" over the next century due to their lack of knowledge about Earth's ice

- There are 5-6 meters worth of sea level in the Greenland ice sheet, and 6-7 meters in the West Antarctic Ice Sheet (some recent studies suggest that the East Antarctic is also melting at alarming rates). Hundreds of millions of people live within that range of sea level increase, so our inability to predict what sea level rise is likely over the next century has substantial human and economic ramifications

- Some recent scientific studies published in the journal *Nature*, however, show that even East Antarctica has been warming by 0.17°C per decade over the past 50 years which is concerning. Antarctica holds enough ice to raise global sea level by 57 m !

- Our lack of understanding of ice dynamics makes predictions of sea level rise uncertain. It is however likely that the current predictions are underestimating the rise to come

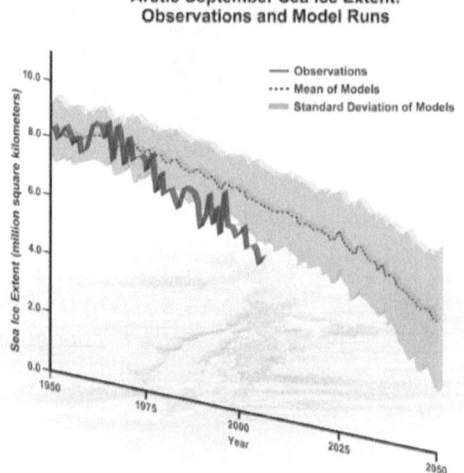

As indicated in this graphic, current observations are already much worse then those predicted by the models

At the current rate, the Arctic will be completely ice free within the decade

Source: IPCC

Slowing Down of the Global Conveyer Belt and The Thermohaline System

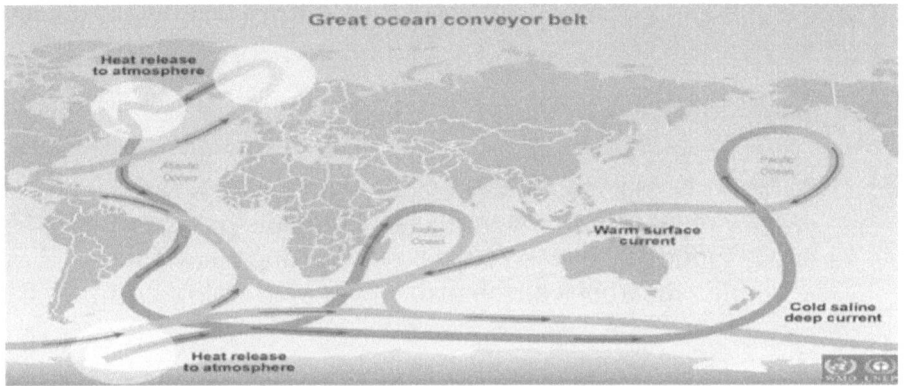

- Most of the weather and climate on the planet is governed by oceanic currents rather than atmospheric circulation

- The thermohaline oceanic circulation is a broad network of oceanic currents which transfers heat around the planet from the poles to the tropics

- The thermohaline circulation is initiated in the Arctic, where particular conditions produce large amounts of very salty water

- The salty water, much denser then the surrounding water, sinks right to the ocean floor, working as a pump mechanism and initiating the global oceanic conveyor belt

- However, Scientists are becoming more and more concerned that due to global warming, the sea water in the Arctic peninsula is becoming less and less salty due to the melting of land and sea ice

- If the water becomes too fresh, this sinking trend may slow down to the point of stopping. If this would happen it would result in large scale and unprecedented, rapid perturbations of the earth's climate system

Ozone and Climate Change

There are two distinct environmental problems associated with ozone: ozone depletion in the Stratosphere (upper atmosphere) and pollution from Tropospheric ozone at ground level. Both will be affected to some extent by climate change:

Stratospheric Ozone:

- While industrial products like chlorofluorocarbons (CFCs) are largely responsible for current ozone depletion, a recent NASA study found that by the 2030s climate change may surpass chlorofluorocarbons as the main driver of overall ozone loss.

- Ozone thinning can occur when increased emissions of methane get transformed into water in the stratosphere. At high altitudes, water vapor can be broken down into molecules that destroy ozone.

- Climate change from greenhouse gases can also affect ozone by heating the lower stratosphere where most of the ozone exists. When the lower stratosphere heats up, chemical reactions speed up, and ozone gets depleted.

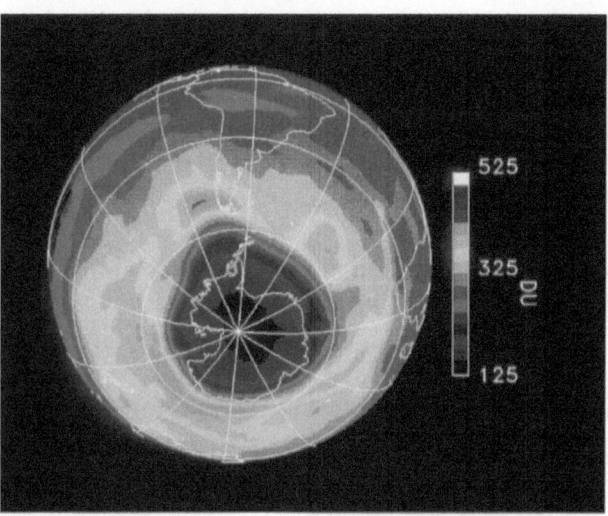

Depletion of Tropospheric ozone over the Antarctic

Case Study: How overconfidence in a solution can give way to new threats

Significant improvements have been made to slow the trend of **ozone** depletion, yet the problem is still with us and we cannot afford to ignore it.

Awareness of the ozone problem has completely lost momentum and seems to have been forgotten by the media. Indeed, we often talk about the ozone problem as if it were a thing of the past – a problem solved. Unfortunately that is not true; the reality is that, while the causes have been identified and international measures taken to limit the trend, ozone related problems and their very real consequences are still present today.

The problems
There are three distinct problems with ozone which often lead to confusion:

The ozone hole: refers to a periodic, near-total depletion of the ozone layer in the upper atmosphere. This depletion occurs every year during springtime over the poles, particularly the South Pole, and their surroundings;

Global depletion of the ozone layer: refers to a slow global thinning of the ozone layer (to a much lesser degree than the ozone hole);

Ground level ozone pollution: is often associated with smog events in urban areas. Ground level ozone forms in a very different way and has nothing to do with the ozone layer in the upper atmospheric ozone.

Of the three issues above, the ozone hole has been the focus of the most attention since its discovery in 1985 during routine atmospheric studies in Antarctica. News quickly spread throughout the world scientific community, which soon identified the ozone hole as a serious global threat requiring an immediate response to avoid a dramatic deterioration of our atmosphere protective layer.

The ozone layer plays a crucial role on Earth by blocking a certain wavelength of hazardous radiations, such as UVBs. Without the ozone layer, life on Earth could simply not exist.

The causes and response

After numerous studies and investigations, scientists determined that human activities were the direct cause of ozone layer depletion. They identified a number of ozone depleting substances (ODS), including chlorofluorocarbons (CFCs), Halons and Freons, which were widely used in the chemical industry and in refrigerator coolants and sprays.

The solution to the ozone crisis was to ban certain chemicals and to replace them with less damaging alternative options. The Montreal Protocol has mostly succeeded in doing this by banning the production of the most harmful ODS.

This response has averted the catastrophic ozone depletion we would otherwise be facing today.

The current situation

The global community has limited the spread of ozone depletion; however, it has not completely stopped the trend. Indeed, two points need to be highlighted:

ODS stays in the atmosphere for a very long time. Therefore, whatever we added in the past (and still do today to a lesser extent) will remain there for decades, or centuries depending on the substances, and will continue to drive ozone depletion for many more years;

While CFCs were banned in most countries (some still use them), the main replacements - hydrochlorofluorocarbons - also contribute to ozone depletion, although to a lesser extent. Therefore, we have in fact only significantly slowed down the trend without really fixing the problem.

The ozone problem is still very present. In fact, scientists recorded the largest ozone hole on record in 2006.

The uncertainties and concerns

Currently, scientists predict the ozone hole will only recover to pre-industrial levels around the year 2065.

Indeed, recent studies indicate that certain parts of the Earth, including major cities, are experiencing dangerous levels of UVBs resulting from global thinning of the ozone layer and the spread of the ozone hole. Higher exposure to this harmful radiation brings significantly higher risks of skin cancer and other chronic illnesses.

Furthermore, while the most harmful ozone depleting substances have been banned, other ODS substances that are not covered by the Montreal Protocol are on the rise, particularly Nitrous Oxides (NOx).

As with climate change predictions, current ozone models do not truly take into consideration the effects of social trends, such as global population growth and the resulting exponential increases in consumerism and chemical activities. As such, current recovery estimates may well be faulty, particularly given the fact that we still do not fully understand the relationship between climate change and ozone depletion.

It may well take much longer than anticipated to resolve the ozone problems, which may in fact worsen with trends in society developments. Meanwhile, for the next 50 to 100 years, harmful effects from damage to the ozone layer – a global health concern - will continue to occur.

Moreover, the ozone layer may face new threats from the rising development of space tourism and the booming aviation sector. The world narrowly averted a similar risk a few decades ago when plans for the deployment of a large scale fleet of supersonic flights, in the form of the Concorde programme, were scrapped. Such activity would have been disastrous to the ozone layer as these planes fly right through it and drive depletion as a side effect of engine combustion.

The key message here is that our atmosphere is fragile, and we cannot afford to destroy it. The ozone problem, followed by the climate change crisis, was our first major warning. Yet we do not seem to have fully learnt the lessons from these on-going threats. We must raise awareness of the continued urgency of the problem, and tread with care in future developments that directly impact our atmosphere.

Tropospheric Ozone:

- According to a recent scientific study by the Royal Society, hundreds more people could die because of increasing levels of ozone at street level. They found that ground levels of ozone, the pollutant created when sunlight hits a mixture of gases in the air (Nitrous and Sulfur oxides), has risen by six per cent per decade since the 1980s.

- Furthermore, tropospheric ozone is known to damage plants, reducing plant primary productivity and crop yields which will have negative impacts on agricultural systems.

- *Depletion of air quality from ozone pollution as a result of increase intensity of photochemical smog from rising temperature is a growing problem*

Urbanisation And Climate Change

- The Addition of GHG into the Atmosphere remains by far the main contributor of man made climate change.

- However, humans are also affecting the climate by modifying the structure of the land through urbanization

- Construction materials (roads, buildings, etc…) absorb much more heat than the natural landmarks (grass, forests, land…) which in turn results in significant local temperature differences

- As our planet is becoming increasingly more and more populated and urbanized, it is not thoroughly understood what the impacts of many local changes could have on the overall climate system

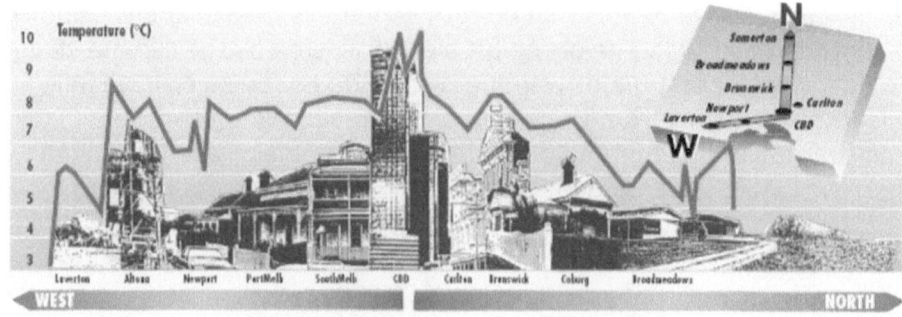

This graphic represents temperatures across Melbourne on a clear and still night. The temperature is warmer in the centre of the city where the amount of constructions and roads is more dense

This phenomenon is known as the 'urban heat island effect' and is another demonstration of how humans actually have an impact on the climate

Land Use And Climate Change

- Before the industrial revolution, much of the earth's surface was covered by forests

- However, in order to feed a rapidly growing population and to satisfy demand for various industrial needs and housing, very significant land surface areas are completely modified by man

- Broadscale changes in land-use patterns, such as deforestation, can significantly alter the roughness and reflectivity of the surface to solar radiation, and hence the absorbed radiation, evaporation and evapotranspiration. In the process, changes in regional climate can occur

- Large scale changes in land use also impact the global climate by enhancing the natural greenhouse effect, are diverse and very difficult to predict; for example by reducing the land's capacity to absorb carbon dioxide (e.g. through deforestation) and by increasing the carbon emission from the land (e.g. through increased biomass decay), both of which lead to greater concentrations of greenhouse gases

Global Dimming

- Since the Industrial Revolution, humans are adding tremendous amounts of particles into the atmosphere (e.g. traces left from jet planes, industrial dust, aerosols)

- These particles, which are not necessarily GHG in nature have a cooling effect by blanketing the lower atmosphere

- The temperature that we are actually experiencing may therefore be cooled down due to the effect of this artificial temporary protective layer.

- Scientists are concerned that a sudden shift in the amount of dimming may result in a rapid increase in surface temperatures.

- Some argue that the dimming of the atmosphere is actually a good thing and therefore that we should not take actions to reduce the emissions of pollutants in the air. This is a missconception that would have severe consequences as it would make the climate unstable and even more subject to extremely rapid changes.

Case Study: Making the Invisible Visible

If carbon dioxide real time emissions were visible to everyone, society would come to realize to a much greater extent what the source of the climate change problem is and in so doing, accelerate the current path to a low carbon economy.

The year 2013 was synonymous with an important date in recent climate change history: for the first time in at least 2 million years, atmospheric levels of carbon dioxide (currently still the main contributor of global warming) have passed 400 parts per million in atmospheric concentrations.

Despite this is only a symbolic number, it was a reminder of the fact that we are on a dangerous path and that our efforts to significantly reduce global carbon emissions have failed. Since then, some significant changes have been made towards a faster deployment of renewables and cleaner technologies but the global economy is still to date very much fossil fuel based with still very impactful consequences on the global climate system...

A common misconception

It is a common misconception to associate fumes visible to the human eye with carbon dioxide emissions. This is false as carbon dioxide is in fact an invisible gas to the human eye. Fumes coming out of factories and commonly portrayed in the press to illustrate global warming are often nothing more than water vapor mixed with other trace substances.

Media often portray carbon dioxide emissions with such images which are an inaccurate representation of what is really happening and gives the false perception that carbon emissions only come from certain punctual sources...

If we could see carbon dioxide in real time, we would realize that carbon dioxide is emitted from multiple sources spreading and mixing around, forming a fog-like mass before creating certain clusters as the gas elevates into the atmosphere. We would also see clearly higher concentrations of the gas at certain specific sources like factory chimneys but also many others less expected sources like above highways

An attempt to make this visualization possible was recently carried out However much improvement would need to be made for mainstream application. A rare camera was developed by a company and surprisingly as a simple tool to help businesses identify CO2 leaks within their operations, perhaps they have not realized the potential that this technology could play for a much larger and important cause...

This misconception, although often unintentional, has had very negative implications because as a result people tend to associate visible fumes with the level of carbon dioxide being emitted into the atmosphere. In fact the pure carbon dioxide emitted in large quantities is omnipresent and invisible around us. Furthermore, CO2 often originates from sources we would not suspect are the biggest emitters.

Making the invisible visible

We don't tend to react fast enough to things that we don't see. Our brain is not meant to react to invisible threats especially when these threats build-up over long periods of time.

In theory, it would be technologically possible to isolate carbon dioxide from other emissions and to visually display the output. Such processes are commonly used by astrophysicists to identify the composition of gases in stars many light years away, each molecule having specific wavelength signatures.

However, the process is time consuming and requires modelling and image processing. If we could improve this technology and easily display carbon dioxide emissions on a real time video or a photographic media, it would revolutionize the way we view the climate change problem and possibly how we react to the threat.

In addition to light analysis, there are already various chemical processes which can make CO2 visible. However, if the idea is to make CO2 emissions widely visible, a chemical reaction would not be feasible on a large scale.

While some work has been done to try to visually showcase carbon emissions, these remain mostly reconstitutions and not real time images. Another good example (in addition to a recent satellite imagery work by NASA) has been the reconstitution to showcase the carbon emissions of New York City.

Business implications

If such technology was made widely available, perhaps a simple solution to reducing global carbon emissions would be to spread this carbon emissions visibility. In a similar way that sustainability reporting is becoming more and more of a practice; emissions visibility disclosure could play a key role in accelerating a shift to a low carbon economy.

Being able to see where most emissions come from in real time would also allow better management and control.

It is foreseen that many businesses would rather not display their emissions and are likely to oppose the idea. However, people and society should really ask themselves if we are still in a position to delay much longer?

The reality is that climate change predictions are rather grim and that we are already heading towards worst case scenarios which will become even worse if we continue business as usual.

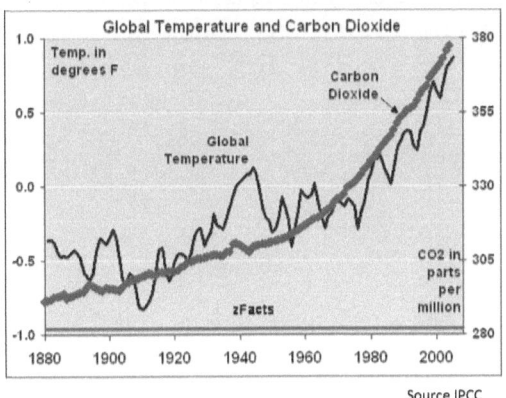

Source IPCC

There is an abundance of data and graphic outputs from multiple reliable independent sources (e.g. NASA, IPCC, Research Institutes, Universities...) which all clearly and consistently show the strong correlation between atmospheric carbon dioxide concentrations and the global atmospheric average temperatures, yet even such strong message representation does not seem to be good enough to drive the necessary level of change...

We only have a matter of years to significantly shift to a low carbon economy. While changes will happen anyhow due to the already present levels of greenhouse gases that we have added and the lag in response of Earth systems, we are still in a position to significantly reduce the severity of the impacts such as sea level rise.

Such visualization disclosure approach would raise awareness on the source of the problem more effectively; pressure society and decisions makers and private sector; allow transparent monitoring of CO_2 emissions at different levels (city, country, targeted locations).

The move to a greener economy is achievable with technologies today. What is lacking is a trigger to pressure societies to move faster. For this to happen, perhaps drastic measures such as this one is needed.

On a last note, the low carbon economy is a tremendous business opportunity, one that business should not fear but on the contrary embrace as an imminent safe growth strategy. Nothing will prevent this change from happening, which obviously is already very much under way, it is just that some more than others are having a hard time moving away from their hold and outdated business models...

VIII. A case study:
CLIMATE CHANGE IN SINGAPORE

- With its 2012 Green Plan and Climate Change strategy, Singapore is starting to take much needed climate change mitigation and prevention measures.

- As an Island Nation Singapore's vulnerability to the effects of climate change is considerable when compared to continental nations.

How does Singapore fit in the global picture?

Carbon Dioxide Emissions from Fossil Fuel Burning Per Person for Top Ten Countries and World, 2006

Country	Emissions Per Person Tons of Carbon
Qatar	22.4
United Arab Emirates	13.3
Kuwait	10.4
Singapore	9.2
USA	5.5
Canada	5.4
Norway	5.3
Australia	4.5
Kazakhstan	4.1
Saudi Arabia	3.9
World Average	**1.3**

Source: Compiled by Earth Policy Institute

Singapore has amongst the **worlds largest CO_2 emissions per capita**

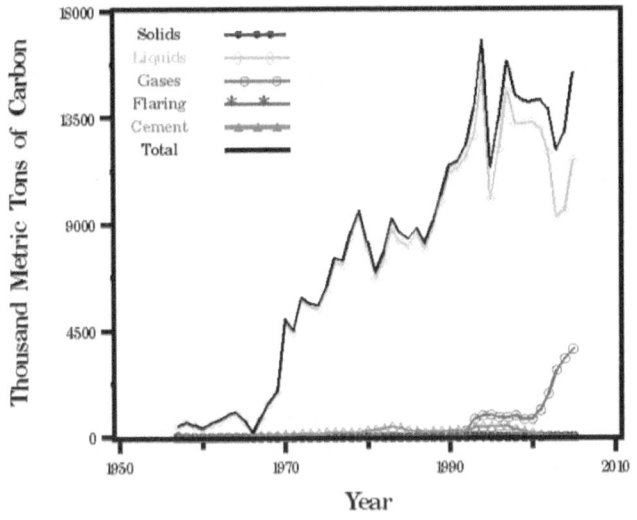

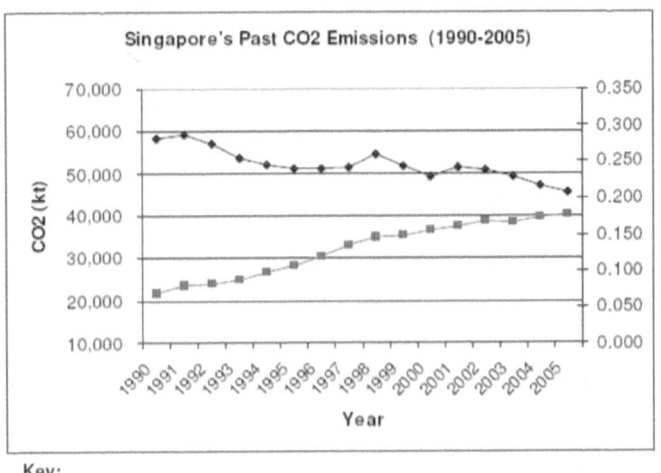

Source MEWR

Singapore CO_2 Emissions

- In Singapore, carbon emissions have sharply increased in the 70's following industrial development.

- The emissions from liquid petroleum have slowly decreased in recent years due to a shift from a petroleum based energy source to natural gas.

- However, the exponential increase of cars on the roads in Singapore is resulting in a further increase of liquid emissions.

Impacts of Climate Change in Singapore: The Risks

Singapore will not be spared from the effects of climate change, the main ones are:

- Increased frequency and intensity of rainfall and tropical storms

- Higher risk of coastal erosion and flooding as a result of rising sea level

- Loss of water resources due to a hotter, drier weather and/or contamination of reservoirs due to saltwater intrusion

- Risk of food shortages due to increased drought due to shifting rain patterns and reduction in rainfall

Singapore will not be spared from the effects of climate change, the main ones are:

- Resurgence/increase of tropical diseases including dengue fever

- Higher rate of reccurrence of the annual haze blanketing the region since 1997

- Loss of local biodiversity

- Economical impacts on Singapore as a result of global crisis in the region and worldwide

- Long term safety issues from a rise in environmental refugees in the region

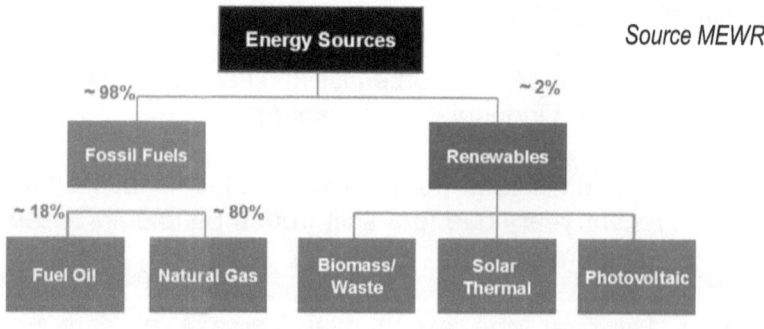

Source MEWR

Singapore Energy sources

- Singapore's energy intensity improved by 15% between 1990 and 2005 due to the adoption of more efficient technologies in power generation and the more productive use of energy in other sectors

- However, Singapore has still a long way to go in providing a sustainable source of energy. Currently only less than 2 % of the energy supply is from renewable source.

	Electricity Generation	Industry	Transport	Buildings	Consumers/ Households	Others
Primary Consumption (combust fuel)	19,315 (48%)	13,465 (33%)	7,056 (17%)	325 (1%)	216 (1%)	-
Secondary Consumption (use electricity)		8,328 (21%)	930 (2%)	5,910 (15%)	3,415 (8%)	732 (2%)
Overall		21,793 (54%)	7,986 (19%)	6,235 (16%)	3,631 (9%)	732 (2%)

TOTAL CO_2 = 40,377 kilo tonnes

Source MEWR

The Singapore context

Facts and figures:

- Despite being a very small nation, Singapore has amongst the **world's largest CO_2 emissions per capita** (about 9 times higher than the world average!).

- Singapore has a very high average energy consumption per capita with **3403 kg oil equivalent.**

- In Singapore most of the energy is generated from natural gas with significant GHG emissions involved

- In Singapore most of the waste is incinerated which is also a major source of GHG emissions.

- Much of Singapore's economy relies on fossil fuels. Singapore is home to major refineries and has one of the busiest ports in the world.

- Singaporeans have built a lifestyle based on the excessive use of air conditioning, which is a major contributor of high electricity consumption and associated GHG.

- Singapore has been a signatory to the UN Framework Convention on Climate Change (UNFCCC) since **1997.**

- Singapore ratified the Kyoto protocol in **April 2006**

Impacts Of Climate Change In Singapore: Water Resources

- Singapore has limited fresh water supplies

- Despite predictions of increased rainfall, rising temperatures will mean more loss through evaporation

- Rising sea level could cause salty intrusions into the water supplies (both above and below ground) and contaminates them

- Flooding may become a problem in some areas of the city. Singapore has however already started to take active measures to limit such disasters with initiatives such as the Marina Barrage which among other benefits helps to reduce flood risks.

As an island nation, sea level rise is a major threat to Singapore

Impacts Of Climate Change In Singapore: Sea Level Rise And Coastal Erosion

- Surrounded by sea and almost entirely flat, Singapore is without doubt vulnerable to the rising sea levels

- A substantial amount of land has been 'reclaimed' from the sea, and these areas are particularly vulnerable to rising sea levels

- Singapore highest point is a hill rising 165 meters above sea level. Most of the areas of commercial and strategic importance of Singapore - its airport, business district and its busy container ports, lie less than two meters above sea level.

- Coastal erosion as a consequence of sea level rise will become a problem for Singapore

- Recorded past extreme high tide events provide an idea of the kind of threat that Singapore will have to deal with

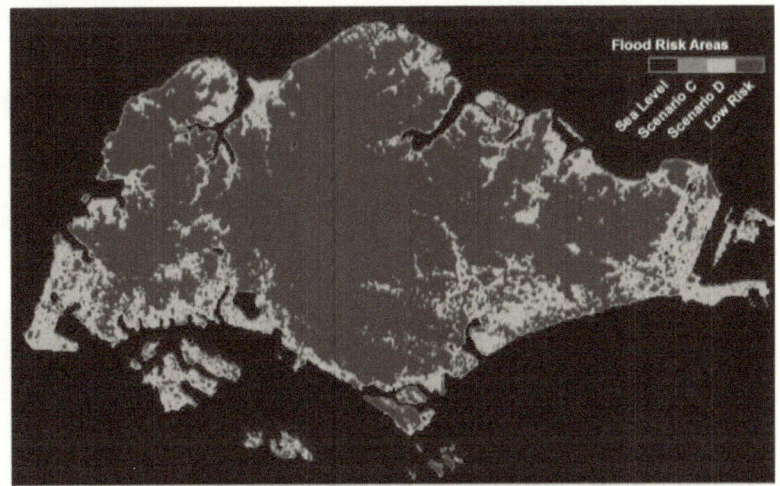

Source NUS

Scenario	Predictions	Map Timescale Description
A	IPCC AR3 lower margin	9cm increase in global MSL by 2100
B	IPCC AR3 upper margin	88cm increase in global MSL by 2100
C	IPCC upper margin + 99/00 NASA JPL La Nina upper margin	+32cm increase in regional sea level to the previous projection
D	ACIA scenario of Greenland icesheet melting	7 metres rise in global MSL

- Much of Singapore's water supply relies on reservoirs (Photo: Lower Peirce Reservoir)

- Salt water intrusion as a result of sea level rise may threaten these reservoirs

This is what would happen to the landscape of Singapore by the end of the century if sea levels were to rise by 88cm - a very likely scenario

Source: NUS

Area loss:

4.02 km2 by 2050
8.11 km2 by 2100

Impacts of Climate Change in Singapore: increase of infectious diseases

- One of the serious consequences of climate change in Singapore will be a significant increase in the number of vector born diseases

- With rising temperatures, more humidity and more rain the conditions will become more conducive for mosquitoes to develop

- Singapore is already experiencing a sharp increase in the number of reported Dengue Fever cases

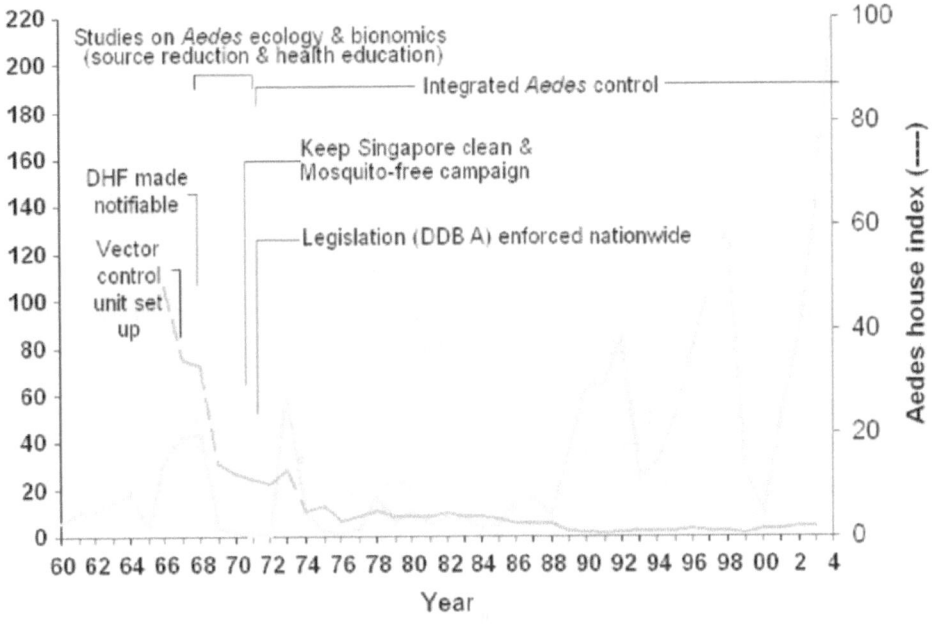

- Impacts Of Climate Change In Singapore (3): Increase Of Infectious Diseases

Impacts Of Climate Change In Singapore: Air Quality

- As temperature rise, the frequency and severity of haze events as a result of more frequent fires in nearby countries is expected to significantly increase

- Air quality will also be affected through an increase in the severity and frequency of smog events. These are caused by a combination of increased temperatures and increased emissions from cars.

- Ozone related deaths and repiratory diseases are expected to significantly increase due to deteriorating air quality.

- Children and the elderly are likely to be most affected by deteriorating air quality.

Impacts of Climate Change in Singapore: biodiversity loss

- Singapore is home to numerous species of endemic plants and animals

- Many of these organisms will be at increased risk of extinction solely as a result of global warming which will add on to the already intense pressures from important development and pollution in Singapore

- In an attempt to protect species at risk, a red list of endangered species in Singapore has been put in place

- The overall aim of the Red List is to convey the urgency and scale of conservation problems to the public and policy makers, and to motivate the global community to try to reduce species extinctions.

- Singapore will not be spared from important biodiversity losses and changes as a result of climate change

- Singapore is home to extensive mangrove ecosystems, which will be threatened by rising sea level and temperatures

- Sungei Buloh wetland reserve, an important site with rich biodiversity, which is a platform for migratory birds

Singapore's Economy and Climate Change

- Singapore will not be spared from the economical impacts of climate change

- As weather patterns become erratic, world food production will be affected which will in turn affect the cost of food. As an island state, Singapore imports most of its goods and is very vulnerable to variations in world market prices

- For instance If a drought or floods affect countries which imports food products to Singapore this will significantly and directly affect the cost of food in Singapore.

- As temperatures rise, natural fresh water supplies will diminish. The cost of producing fresh water from various sources will rise (e.g. desalinisation, reverse osmosis, etc…).

- The cost of efforts to mitigate the harmful effects of sea level rises and other climate change related impacts (e.g. biodiversity, control of infectious diseases…) will weigh heavily on Singapore's economy

Singapore National Climate Change Strategy

- The Singapore National Climate Change Strategy underscores Singapore's commitment to do its part in the international effort to address climate change. The strategy includes plans to mitigate greenhouse gas emissions. It also proposes efforts to better understand Singapore's vulnerabilities to climate change and to assess adaptation measures to address the impacts of climate change

- The strategy identifies improving **energy efficiency** as Singapore's key strategy for mitigating greenhouse gas emissions.

- The Singapore government will actively support energy users in the industry, buildings, households and transport sectors to be more energy efficient, through regulations to provide consumer information and deploy appropriate technologies, and incentives to encourage take-up among individuals and businesses.

- The government will also encourage further research and development into clean energy.

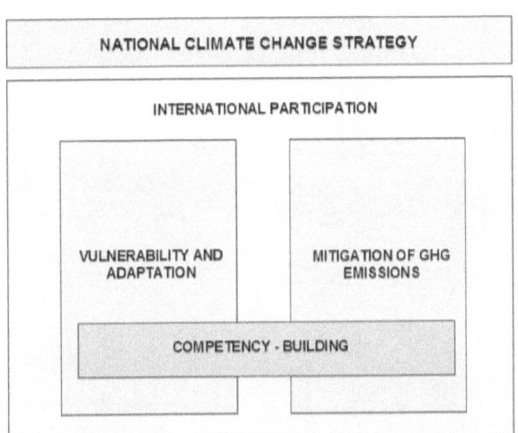

Source MEWR

Measures Taken in Singapore to Minimize the Impacts of Climate Change

In many ways, Singapore has already implemented several measures that are useful for cities across the world to model their climate change mitigation plans on.

Water Planning:

- In 2008 Singapore opened the Marina Barrage which is a unique example of water management in urban areas

- The barrage, coupled with a state-of-the-art desalination plant and a sewage treatment plant provide Singapore with a water sustainable supply.

City in the Garden concept:

- With its concept of a Garden City, Singapore has remarkably preserved the trees within the city

- Singapore has also dedicated a significant part of its land area for the preservation of nature reserves and gardens

Green Buildings:

- The Building and Construction Authority has launched its Green Mark standard which rewards environmentally friendly energy efficient buildings

IX. WHAT CAN BE DONE? SOLUTIONS TO FIGHT CLIMATE CHANGE

Carbon Footprint

In order to successfully fight climate change, governments, businesses and individuals must collectively resolve to reduce their carbon footprint

Where are actions needed?

In order to win the fight against climate change, efforts must be undertaken in the following areas:

- Develop technologies to remove GHG from the atmosphere

- Develop technologies which may help to cool the planet (artificial dimming)

- Prepare to adapt to a changing environment

- Research on the uncertain mechanisms of climate change (feedback effects, ice dynamic etc...)

- Closely monitor the evolution of climate change impacts

- Educate the population on the risks and contributions of their activities to climate change, and on how to reduce their individual ecological footprints.

What can Individuals do?

- Practice responsible consumerism by purchasing products which are energy efficient and have a low carbon footprint.

- Choose energy efficient modes of transportation (public transport or purchase fuel efficient vehicles such as hybrid cars).

- Practice an energy efficient life style at home and at work (e.g. switching off lights in unoccupied rooms, raising air conditioning temperature to 25 °C...).

- Stay informed and spread awareness on climate change issues

- Get involved by helping, participating in environmental groupsworking towards mitigating climate change.

- Vote for political leaders who are taking climate change issues seriously.

Climate change is a threat to world peace and stability and should be taken very seriously

The decisions that we make today will affect not only ourselves but most importantlythe environment in which future generations will have to live in

What can corporations do?

- Corporations have the responsibility to conduct their business in a sustainable way

- Purchase and produce environment friendly products which are energy efficient and have a low carbon footprint

- Construct environmentally friendly buildings and office space which have a low carbon footprint

- Implement environment management systems and ecofriendly practices within the workforce

- Report environmental impacts to the community (CSR)

- Educate staff on good environmental behavior and on climate change issues

- Support environmental movements and groups which are active in the fight against climate change

Examples of leading reducers corporations

There are now numerous examples across the world of very successful corporation's which have achieved impressive savings by adopting sustainable practices through Energy/water savings and more generally by adopting sustainable development models.

NORSKE CANADA reduced emissions 61% on 1990 levels and saved estimated CA$20-30m

IBM reduced emissions 65% on 1990 levels and saved US$791 million

BRITISH TELECOM reduced energy-related emissions 62% on 1992 levels and saved £119 million (US$200m)

DUPONT reduced greenhouse gas emissions 69% from 1990 levels and saved US$2 billion

BP reduced emissions 18% from 1997 to 2001, saving an estimated US$650m on a $20m investment

...

What can governments do?

- Political will is a major driver for the fight against climate change

- Governments around the world must take action and adopt policies which facilitate the shift from a fossil fuel economy to a low carbon future

- Governments must facilitate the implementation of renewable energy and sustainable transportation modes

- They must create green jobs to make the transition to a green economy feasible

- Collaborate with other Governments in solving the climate change crisis. climate change is a global issue that can only be solved through international collaboration

- Adopt and follow international agreements on mitigating GHG emissions (Kyoto protocol, agenda 21)

- Developed countries must help developing countries reduce their carbon emissions by facilitating the transfer of renewable technologies and practices to these countries

- Simple technologies such as affordable energy efficient stoves and wind turbines can help make a difference if applied on a large scale

International Agreements: The Kyoto Protocol

- The Kyoto Protocol is an international agreement linked to the United Nations Framework Convention on Climate Change. A key feature of the Kyoto Protocol is that it sets binding targets for 37 industrialized countries and the European community for reducing Greenhouse Gas (GHG) emissions

- These amount to an average of 5 %t against 1990 levels over the five-year period 2008-2012.

- Countries have a certain degree of flexibility in how they make and measure their emissions reductions

- They pursue emissions cuts in a wide range of economic sectors

- The Protocol advances the implementation of existing commitments by all countries

- The EU and its Member States ratified the Kyoto Protocol in late May 2002

- To date 175 countries in total have ratified the Kyoto protocol, the latest being Australia in 2007

- n 2008, the USA remains the last developed country that refuses to ratify the protocol

- The Protocol advances the implementation of existing commitments by all countries

- The EU and its Member States ratified the Kyoto Protocol in late May 2002

- To date 175 countries in total have ratified the Kyoto protocol, the latest being Australia in 2007

- n 2008, the USA remains the last developed country that refuses to ratify the protocol

International Agreements: Beyond Kyoto

- The Kyoto protocol is now in its active phase (2008 to 2012) in which countries who have ratified the protocol are actively undertaking measures to significantly reduce their carbon emissions

- However, governments must start planning beyond Kyoto and put into place long term plans which will progressively allow a transition from a Carbon based economy to a sustainable one

- One must understand that the Kyoto targets are mostly symbolic in having nations all over the world come together for the cause of fighting climate change. Climate change is a global matter which requires global collaboration. However, much stronger measures than those outlined by the Kyoto protocol are required to be effective. This is why nations must now start to seriously think at longer perspective and stronger measures.

- Climate talks took place in Poznan in December 2008 to discuss such post-Kyoto issues

Shifting from a fossil fuel based economy to a low carbon future is imperative if we are to succeed in fighting climate change

Geothermal Energy

- Geothermal energy is the process of using the heat from the Earth to produce steam and generate electricity in the process

- Geothermal resources range from shallow ground to hot water and rock several miles below the Earth's surface, and even further down to the extremely hot molten rock called magma. Wells over a mile deep can be drilled into underground reservoirs to tap steam and very hot water that can be brought to the surface for use in a variety of applications.

- At present, geothermal energy is only used in areas where it is easy to have access to the steam such as in volcanic active places

- As of 2008, geothermal power supplies less than 1% of the world's energy

- However, new developments in drilling technologies will soon enable drilling deeper into the ground and access to such resources in almost any places around the world

- Geothermal has the potential to provide unlimited clean energy for generations to come

Solar Energy: Photovoltaic

- There are two ways in which solar energy (radiation from the sun) can be utilised as Energy sources:

1. Photovoltaic Energy: which is the process of converting solar radiation into electricity through solar panels

- Photovoltaic production has been doubling every two years, increasing by an average of 48 % each year since 2002, making it the world's fastest-growing energy technology. World solar photovoltaic (PV) market installations reached a record high of 2.8 gigawatts peak (GWp) in 2007

- Although the selling price of modules is still too high to compete with grid electricity in most places, the rapid production growth is expected to substantially cut down the price, which will make photovoltaic energy competitive within a few years

- Currently average Sunpower's cells have a conversion ratio of 23.4 %. However, advances past this efficiency mark are being innovated and efficiencies of 42% have been achieved

- New technologies of photovoltaic cells are expected to hit the market soon: Fine layer cells and dye cells which will make solar cells much easier to integrate into buildings

- New research are under way to capture solar energy directly from space and to transfer it down to earth via micro waves

- Concentrating solar energy into new generation high temperature resistant solar cells is also under way and already experimented in Australia. Such technology has great potential

- Solar photovoltaic energy is one of the most promising renewable energy sources which may well end up powering cities of the future

Solar Energy: Thermal

2. **Solar thermal energy (STE)** is a technology for harnessing solar energy for thermal energy (heat)

- Low temperature collectors are flat plates generally used to heat swimming pools

- Medium-temperature collectors are also usually flat plates but are used for heating water for residential and commercial use.

- High temperature collectors concentrate sunlight using mirrors or lenses and are generally used for electric power production. This is different from solar photovoltaics, which convert solar energy directly into electricity

- The principal behind high temperature collectors is to concentrate the heat in order to produce, which powers turbines and generates electricity

- This process is quite efficient in countries which have large surface areas and appropriate weather conditions (sun exposure)

Wind Energy

- Wind power is the conversion of wind energy into electricity, using wind turbines

- At the end of 2007, the worldwide capacity of wind-powered generators was 94.1 gigawatts. Although wind produces only about 1% of world-wide electricity use, it is growing rapidly, increasing more than fivefold globally between 2000 and 2007

-

Case Study: The Energy Dilemna

Nuclear energy is a possible but potentially dangerous solution to the need for low-carbon growth. The renewable energy sector must grow rapidly to reduce global reliance on nuclear

While opposition to the deployment of nuclear in the developing world is a hot topic, this energy source is consistently topping the list as the short term solution to the climate crisis because it is the only one that addresses a critical dilemma. What is this dilemma, and why is a potential large scale deployment of nuclear reactors a concern?

What is the dilemma?

Rising levels of greenhouse gas (GHG) emissions are becoming alarming. Levels of GHG have already passed threshold safety levels and we are now at the start of a red zone of socio-environmental consequences, such as natural disasters and resource shortages. These will only intensify as GHG atmospheric levels keep rising. In order to keep the rising threats of climate change under control, GHG levels in the atmosphere should be stabilized.

Secondly, the reality is that global energy demand is rising very fast, especially in the developing world. Many countries are still opting for cheaper fossil fuels to meet this growing demand. Because of economic pressures, high levels of energy production are required to keep up with countries' development targets.

The fact is, we have to cut the world's CO2 emissions in half by 2050. However, energy demand, a major source of CO2 emissions, will double by that date. Thus, one of the greatest dilemmas of our time is defined as meeting the rise in energy demand while at the same time addressing the climate change threat through cutting down GHG emissions.

Renewables and the energy demand

In order to achieve the GHG reduction targets, many governments have opted for renewable energy. While the idea is promising, and solar and wind energy capacities have increased significantly recently, the deployment of renewables is not yet coping with the rise in energy demand.

The global energy mix is still very strongly dominated by fossil fuels. Coal and natural gas are likely to remain the main sources of energy in the next few decades with nuclear potentially catching up fast.

For renewables to capture a greater share of the global energy mix is possible and a matter of political will. However, the trend is currently moving too slowly to avoid a catastrophic build-up of GHG in the atmosphere.

An unwanted but likely solution driven by the dilemma

This opposition [to nuclear energy] combined with the need to cut down GHG emissions has created an open door for renewables but the move to deploy renewable energies on a large scale should be much faster than it is currently occurring.Nuclear energy seems to be a viable option that would allow us to address the challenges of the dilemma fast enough. However, because nuclear energy is not without major security and environmental concerns, a large scale deployment of renewable energy would be a better alternative.

Looking at the global picture, a few countries are already considering deploying nuclear reactors rather than renewable energy facilities as their primary strategy. Despite a relative slowdown in Europe, China may well be building hundreds of nuclear reactors to meet its newly established emissions targets and the UAE has already started its nuclear program.

The decisions to go nuclear are also facing strong opposition by the international community in the wake of recent incidents such as the disaster at the Fukushima Nuclear Power Plant in Japan in 2011. This opposition combined with the need to cut down GHG emissions has created an open door for renewables but the move to deploy renewable energies on a large scale should be much faster than it is currently occurring.

Managing risks

Nuclear poses serious risks to global safety and security. Despite accidents being very unlikely and infrequent, they have happened in the past and statistically are likely to happen again. The fact is that nuclear power plants should not be deployed in areas which are vulnerable to either natural risks such as earthquakes or political risks such as unstable governments.

Management of nuclear waste is also a concern. While the volume of waste is very small, its hazard level is high in the very long term. The disposal of nuclear products has improved and they are nowadays sent to old salt mines in geologically stable locations; it is a low risk solution. The major risks with nuclear comes from radioactive leaks from unforeseen events, transportation hazards and terrorism, the latter being of most concern.

In an increasingly unstable world where terrorism is a growing threat, nuclear reactors are prime targets that would be difficult and very costly to protect. For instance recently a number of nuclear power plants have been flown over by unidentified drones in France, a major security breach. If a nuclear incident was to occur as a result of unforeseen or terrorist act in a densely populated area like Europe, the consequences would be humanly and financially catastrophic.

Carbon capture as a likely contender

Unless a significant step up of renewables is achieved, the closest current option contending with nuclear in terms of addressing the challenges of the dilemma is carbon capture. Carbon capture is the process of capturing carbon emissions at the source of production, to store it and to inject it back underground. Nuclear fusion (as opposed to nuclear fission discussed above) is a growing hope but unlikely to be operational for many more decades.

Avoiding what could become a risky boom in nuclear energy would require a significant step up of renewables on the marketplace and/or an improvement in carbon capture technology.

- Large scale wind farms are typically connected to the local electric power transmission network, with smaller turbines being used to provide electricity to isolated locations

- Smaller turbines can be used to provide electricity to isolated locations

- Wind energy as a power source is a viable alternative to fossil fuels, as it is plentiful, renewable, widely distributed, clean, and produces lower greenhouse gas emissions

- New research is underway to capture altitude wind which is more constant and would provide a continuous unlimited supply of clean energy

Hydroelectric Energy

- **Hydroelectricity** is electricity generated by hydropower, the production of power through use of the gravitational force of falling or flowing water. It is the most widely used form of renewable energy

- Once a hydroelectric complex is constructed, the project produces no direct waste, and has a considerably different output level of the greenhouse gas carbon dioxide (CO_2) than fossil fuel powered energy plants (significant amounts of GHG are release during the construction when forests are flooded and gas such as methane emitted during the decay of organic matter)

- Worldwide, hydroelectricity supplied an estimated 715,000 MWe in 2005. This was approximately 19% of the world's electricity (up from 16% in 2003), and accounted for over 63% of electricity from renewable sources

- Hydroelectric plants however have significant impacts on biodiversity and are therefore considered my many environmentalists not to be eco-friendly

Nuclear Fusion

- **Nuclear fusion** is the process by which multiple like-charged atomic nuclei join together to form a heavier nucleus. It is accompanied by the release or absorption of energy

- Nuclear fusion occurs naturally in stars. Artificial fusion in human enterprises has also been achieved, although not yet completely controlled

- Fusion, the reaction that produces the sun's energy, is thought to have enormous potential for future power generation because fusion plant operation produces no emissions, fuel sources are potentially abundant, and it produces relatively little (and short-lived) radioactive waste (as opposed to nuclear fission which is the nuclear power currently harnessed)

- High-power reactors such as the ITER (International Thermonuclear Experimental Reactor) are now under construction in France and other countries

- Physicists estimate that exploitable fusion reactors might be operational within 50 years. The main issue remaining to be resolved concerns the storage and movement of the very hot plasma within the reactor and to make the reaction durable (for now the reaction can only be kept running for about 30 seconds due to the extreme heat generated)

- Nuclear fusion is a very promising, clean and sustainable technology which has the potential to power the entire planet by the end of the century

- However, for now this technology is not operational. Until it is, other sustainable alternatives must be considered until it is improved

"Clean" Coal

- This technology involves the removal of a significant part of the CO_2 from the coal through chemical and physical means before it is combusted. By doing so, there is much less carbon dioxide released into the atmosphere (up to 80% reduction in emissions)

- The CO_2 removed is stored and then injected back into the Earth's crust

- This technology is debatable as it relies on one of the most polluting fossil fuel energy sources

- The idea behind such technology is to try to clean as much as possible the coal burning process which is used by hundreds of thousands of power plants worldwide, while more sustainable alternatives are put into place

- This technology should only be used as a temporary solution, as continuing to rely on fossil fuels is a step in the wrong direction. In order to fight climate change, we must completely cease our reliance on fossil fuels and start relying heavily on sustainable energies instead

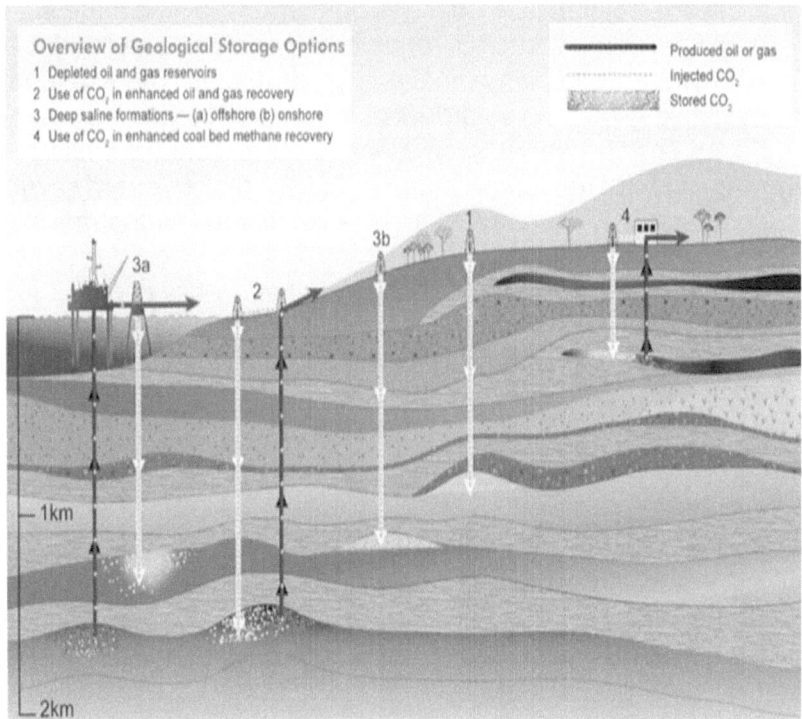

"Clean" coal: Storage of CO_2 in deep underground geological formations

Oceans Thermal Energy

- There is a lot of potential in the utilization of thermal energy from the oceans

- There is an important temperature difference between the bottom of the deep sea and surface temperatures

- The Earth's oceans are continually heated by the sun, and cover nearly 70% of the Earth's surface. This temperature difference contains a vast amount of solar energy which can potentially be harnessed for human use

- Ocean thermal energy conversion (OTEC) is a method of generating electricity which uses the temperature difference that exists between deep and shallow waters to run a heat engine. As with any heat engine, the greatest efficiency and power is produced with the largest temperature difference. This temperature difference generally increases with decreasing latitude, i.e. near the equator, in the tropics

- If this extraction could be made cost effective on a large scale, it could provide a source of renewable energy needed to deal with energy shortages, and other energy problems. The total energy available is one or two orders of magnitude higher than other ocean energy options such as wave power, but the small magnitude of the temperature difference makes energy extraction comparatively difficult and expensive, due to low thermal efficiency

- Earlier OTEC systems had an overall efficiency of only 1 to 3% (the theoretical maximum efficiency lies between 6 and 7%). Current designs under review will operate closer to the theoretical maximum efficiency

Other renewable energy sources

Other sources of renewable energy include:

- **Wave energy** from the ocean

- Hydroelectric Energy from ocean **tides**

Sustainable waste management

- Waste management is an important challenge for the Cities of the future as well as an important contributor to climate change

- The treatment of waste in landfills (incineration and decomposition) is responsible for emitting greenhouse gases such as Carbon Dioxide and Methane

- Improvements must be made in the waste management sector if we are to succeed in significantly cutting down Greenhouse gas emissions

Technologies that must be looked into include:

- Plasma incineration of waste: The extreme heat generated allows to breakdown molecules and generate energy in the process

- Waste to energy: Produce energy from the decomposition or burning of waste

Case Study: Energy Wars, Survival of the Fittest

While the deployment of large scale renewables (mostly solar and wind) has started, the situation is such that without major changes, coal and nuclear are likely to keep a strong global presence in the decades to come. However, there is a strong case for why none of the latter energy sources should remain on a large scale as the solution. In this article is a review of energy options and a clear case for change to a renewable-based economy.

Coal: an unwanted but likely contender

Few economists would deny the era of a petroleum-based economy is coming to an end. Supplies of oil are vanishing fast; despite variations in reported supplies most experts agree that within a few decades "easy" supplies of liquid fossil fuels will vanish. The cost of oil can only go up as the supplies become rare or more difficult to access (e.g. deep sea extraction, tar sands, residual reserves).

However, other non-renewable supplies are still largely abundant, especially coal. While renewables are still (this is about to change fast) struggling to compete financially with the current cost of cheap (and transportable) energy, coal's low financial cost makes it an easy replacement for oil. The world may come to rely on coal once again, and even derive liquid fuels from this resource, as the cost of the process will still be cheaper than that of oil (technology to convert coal into liquid fuel does exist, in fact it was already used during the Second World War).

This trend is already highly visible. Countries like China and the USA (which are likely to remain the two largest economies in the 21st century) are both still investing heavily in coal-based energy. In China alone, new coal-fired power plants still open frequently and almost none of these are oil-based. Furthermore, both of these countries are rich in coal reserves.

The prospect of energy security through a coal-based economy can be seen as a life saver for energy-starved communities, but would really be a time bomb from a climate point of view! As a reminder, 2016 was declared the hottest year in recorded history. The finding highlights the failure of over 20 years of international climate change negotiations. While the debate on climate change is clearly over and warnings of serious consequences of inaction are abundant, we do not seem to be changing course fast enough from what is rapidly becoming an unavoidable catastrophe impacting future (and even current) generations. We have passed the point at which, no matter what we do, significant changes will happen; it is just a matter of how bad these changes will be.

A shift to a coal-based economy would commit us to this self-destructive path for decades to come. With viable renewable technologies available today, there's no excuse for choosing the "easy" coal option.

Nuclear, a viable but risky option

Nuclear fission: Many nations have turned to nuclear energy to cut down GHG emissions to avoid reaching dangerous atmospheric build-up. Nuclear has been a hot debate for many years and despite nuclear's promising start at the end of the Second World War, cheap oil has managed to put the technology in the background, with the notable exception of a few countries like France.

Nuclear would indeed allow nations to significantly reduce their GHG emissions; however there are a few major issues with nuclear fission.

First, most current nuclear reactors rely on fissile uranium (U235 isotope) and the process is highly inefficient. Only about one per cent of the caloric content of the fuel is retrieved while the rest is turned into nuclear waste that will last for thousands of years. At about 30 per cent (less than that of modern coal fired plants), the thermodynamic efficiency of current nuclear plants is rather low as well. If nuclear was to be applied and sustained in the long term, we would need to use advanced nuclear reactors ("breeder reactors") which more effectively enrich uranium into plutonium. These reactors, such as the Super-Phoenix nuclear plant in France, are much more efficient. About 75 per cent of the original uranium is used effectively. However, using plutonium reactors poses serious global security issues as plutonium can also be used to derive nuclear weapons.

Secondly, uranium is not a renewable resource. If uranium were exploited on a large scale (especially for the non-breeder reactor process), power plants would quickly exhaust uranium ore resources (the time scale is about the same as that for oil or gas).

Furthermore, dealing with nuclear waste (transport and disposal) is still a problem, especially if this process was applied on a much larger scale. To date, no country has developed a permanent solution to even the relatively small amounts of nuclear waste generated in today's power plants.

And finally, past and recent accidents involving nuclear plants are putting further doubts on a larger scale implementation of this technology.

Nuclear fusion: Nuclear fusion is seen by many as the Holy Grail of clean energy as it could produce unlimited amounts of energy, mostly from sea water, without nuclear waste and other unwanted consequences. However, the reality is that we are still far from exploiting this technology. Major breakthroughs still need to be achieved to stabilize and confine the reactions. Most experts predict that nuclear fusion will not be commercially available for another 40-50 years. However, we are running out of time. We cannot afford to wait or put all our hopes on fusion.

While it appears that coal and nuclear are the easy solutions to energy security in the coming decades, both would be unwanted outcomes.

The shift from the current fossil fuel economy to a low-carbon economy based on alternative energies is at a tipping point. However, despite the numerous solutions available today; our focus is limited to a few renewable energy technologies. Others remain highly unexploited despite the fact that they could provide a significant contribution. We must continue developing the current primary renewables (solar and wind), but we should also start exploiting other alternatives on a much larger scale than they are used today. No single currently available technology can provide a global solution to the energy crisis.

A better and highly necessary alternative

A better alternative would be a complete shift from fossil fuels to renewables. And although it may seem unrealistic to many, it is indeed possible even based on today's technologies. Here is an overview of the pros and cons of the available options:

The common

Solar photovoltaic: Solar photovoltaic (PV) has greatly improved in efficiency and cost-effectiveness thanks to extensive research and development in recent years. Even so, efficiency remains relatively low (the conversion factor efficiency of commercial photovoltaic panels based on crystalline silicium cells is still on average about 20 per cent) and the cost relatively high. Furthermore, in order to produce significant energy supply from PV, very large surface areas would be required. However, keep in mind the following attractive fact: if we were to convert even a fraction of the world's deserts into solar photovoltaic farms (e.g. the Sahara) we could power the entire planet on solar energy alone. Indeed, the average solar radiation power falling per year on a square meter in the Sahara is higher than 200W

Based on this fact alone, solar photovoltaic is and will remain a major area of focus and hope. New technologies such as concentrated solar and solar films will significantly change the current usage of solar cells. Solar energy could also be harvested directly from space and brought back to Earth in the form of microwaves (such technology is possible and countries such as Japan are considering the option).

As Thomas Edison once said: *"I'd put my money on the sun and solar energy. What a source of power! I hope we don't have to wait till oil and coal run out before we tackle that".*

Wind energy: Wind energy has been one of the most rapidly expanding renewable energy source in recent years and offers great potential. However, we must not forget (and it is often forgotten) that modern wind turbines are complex machines which are resource- intensive to make and maintain. For example, one single turbine contains kilometres of copper wires for which mining and production is a highly polluting and energy-intensive process. From a life cycle point of view, wind turbines actually emit significant amounts of greenhouse gases (GHGs). However, wind turbines do pay off over time from an energy perspective and the potential to exploit wind patterns (especially offshore where they do not compete for space) is promising. The main advantage of wind is that it is widely available.

Modern wind turbines can generate about 100 megawatts (MW) per square kilometre, which is on par with solar thermal. A major disadvantage of wind power is that it requires storage capabilities (the grid can only absorb variable energy output if wind power is limited to a small percentage of total installed power). It is quite realistic to assume that wind power will contribute no more than about five per cent of total power production as a larger implementation would require significant changes to existing electrical networks.

Hydropower dams: Hydropower dams are well used in countries that have access to large amounts of fresh water. Hydropower works well from an energy efficiency point of view but the construction of dams, especially in biodiversity-rich areas, are sometimes truly devastating. Furthermore, the construction of large scale dams (such as the Three Gorges Dam in China) can be responsible for the disappearance of entire villages, cultural heritage sites and cause major disruptions to natural river flows. For these reasons alone, hydropower dams are in many cases far from a sustainable alternative.

In addition, the large amount of cement required, as well as the destruction of forests, are responsible for the release of significant amounts of GHGs. Hydroelectric project should be evaluated carefully for their impacts on both the surrounding environment and the communities. Smaller scale hydroelectric projects sometimes make more sense as the destruction to the surrounding areas is minimal. On a smaller scale, there is significant potential for micro hydroelectric turbines. These systems, which make use of the pressurised flow in transportation pipes and other installations such as buildings, remain virtually unexploited.

The less common

Passive solar (largely unexploited): New solar passive technologies which use the sun's heat to generate steam to spin turbines are on the rise and very promising. Several long term trials of solar towers (concentrating mirrors on a central heat exchanger) have been held in locations such as Seville, Spain or in the United States. Meanwhile, Australia is constructing another type of passive solar technology on a massive scale: tower funnels that will produce energy in the range of 200 MW. Modern solar thermal power plants have a conversion efficiency of about 30 per cent, about the same as nuclear plants. Furthermore, the energy payback time for solar thermal is highly favourable at about 5 months. We are only beginning to realise the potential of this relatively simple technology, which is reliable and requires little maintenance. One drawback is that it requires large surface areas with high solar exposure. However, as with photovoltaic, desert regions exist where such technology could be implemented on a large scale.

Ocean thermal energy (largely unexploited): Ocean thermal energy has great potential and is one of the least exploited renewables. There is a significant thermal gradient, or temperature difference, between the deep ocean and the water surface. Commercial exploitation of this energy would require significant investment for relatively little output, but it can be accessed almost anywhere on an ocean and is a constant source. Furthermore, new technologies suggest that the process efficiency could be significantly improved by using passive solar to increase the temperature gradient between the deep sea and the surface.

Tidal energy: Tidal energy which is constant and very powerful is also largely unexploited to date. Many locations on the planet have particularly strong tidal currents. A range of new turbine technologies have been tested in recent years and proved that the biggest barrier to the energy source – corrosion – can be reduced. Furthermore tidal energy is predictable (as opposed to wind) and therefore reliable.

Geothermal (largely unexploited): If we put the technology difficulties to the side and focus on the problem as a whole, we come up with the following observation: we are on a planet which contains mostly melted rock at very high temperatures and we are only on a small crust which sits on this magma. We can therefore see our planet as a giant "cooker" with an unlimited supply of energy just below our feet! Of course the situation is a bit more complicated when it comes to drilling and accessing this heat. However, the situation has changed dramatically in recent years as advances in drilling technologies (quite ironically mostly developed by oil and gas companies) now allows drilling at much greater depths and makes the exploitation of geothermal energy possible in many parts of the world that were before non-exploitable. Furthermore, an advantage of geothermal is that it does not require the use of much land area, giving it an advantage over more land-intensive renewables that are increasingly competing with agriculture and forests for space. Much more research should be allocated to the development of geothermal.

Bioenergy: Bioenergy has big potential; despite serious mistakes made in the early development of biofuels. Decisions to replace large swathes of agricultural fields with biofuel crops have led to global food price rises and security issues. Furthermore, biofuel crops have been a major driver of deforestation. However, new developments such as algae biofuels, which would cultivate supplies from oceans rather than on land, are promising although still experimental. Bioenergy's main advantage is its emissions reduction potential - recently grown biomass does not contribute to climate change as fossil fuels do. Furthermore, there is enormous opportunity to convert the tremendous amounts of waste generated by society to biofuel (waste to energy).

Others: Several less conventional potential renewable sources have barely reached the level of experimental research. One example would be deriving energy from bacterial activity on a large scale (many bacterial processes are exothermic and the free heat generated could be used to derive energy).

Tidal energy: Tidal energy which is constant and very powerful is also largely unexploited to date. Many locations on the planet have particularly strong tidal

The forgotten energies:

Clearly the emphasis is increasingly on photovoltaic and wind energies when it comes to renewables. We should continue and even accelerate those efforts. But at the same time, other renewables need much greater attention and research if we are to avoid the unwanted outcome of a major shift from petroleum to a coal-based energy.

Geothermal, passive solar, ocean thermal and tidal energies in particular should become predominant focus points. We need to "renew renewables" in the sense that in addition to current photovoltaic and wind, other renewable energy sources deserve their chances of large scale deployment. They may actually prove more successful at providing sustainable alternatives to the energy crisis.

Which energy source do you think is the most fit for success?

Personally, I believe the solution relies in a rapid progressive shift to a mixture of renewables in all their forms. However, I find the single below graphic representation to be particularly striking and to speak for itself. It should be kept in mind by any investors or strategists out there...

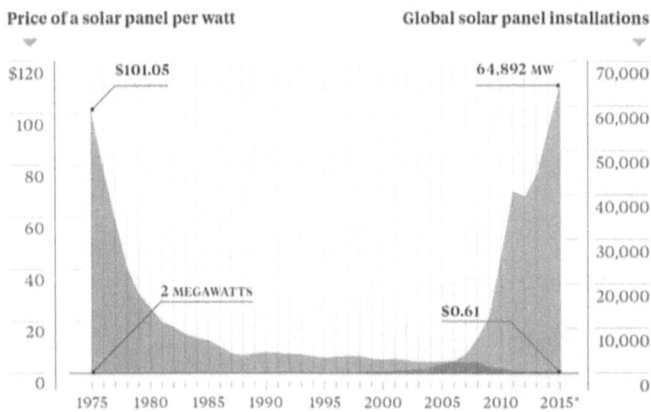

source: Earth Policy Institute

133

Sustainable transport

- Current transportation is largely based on fossil fuels

- The development of more sustainable modes of transportation is essential to reducing GHG emissions

- Alternative fossil fuels such as diesel and natural gas still produce very significant GHG emissions and should only be seen as temporary transition solutions while much cleaner technologies are developed and improved

- The shift from fossil fuel based kerosene to biofuels is slowly on its way in the air travel industry

- Cities should rethink their transportation routes and especially integrate effective and strategically spread walking and cycling tracks to allow easy and convenient displacement within the city. Such installations would also ease the development of new modes of transportations based on human energy

- Entire models of lifestyle and habits need to be seriously rethought. Hundreds of millions of people travel each day to get to and from their workplaces. Is this really necessary? With the technology available today in most cases the work can be done from home through simple communication media (videoconferencing, phone, fax, emails etc…). If we are to significantly reduce carbon emissions such habits that we have considered granted for centuries must be rethought. The same goes for business travel, which is in most cases not necessary considering the available technology

Sustainable modes of transport
Electric vehicles
Fuel cell vehicles (Hydrogen)
Solar vehicles
Sustainable public mass transportation
Water powered vehicles
Compressed air vehicles
Human powered vehicles (bicycles etc….)

Biofuel

- Biofuel is beneficial in cutting down carbon emissions because the carbon emitted has been removed from the environment in present times, as opposed to fossil fuels which release passed long time stored carbon into the atmosphere (Therefore adding to the present GHG concentrations).

In order for biofuel to be a sustainable alternative to fossil fuels a number of important issues must be solved:

- Currently biofuels are extracted from land plantations (e.g. corn, soya, etc.) which uses large tracts of land, encroaching on land used to grow food crops. competing with food crops
- In addition, crops for biofuel is a threat to land biodiversity if it is exploited on a global scale
- Utilising biofuel from aquatic origin appears to be a much more reliable source
- Another disadvantage of biofuel is that it still produces toxic emissions such as nitrous oxide, which causes air quality to deteriorate and generates health problems

Sustainable Buildings

- Buildings are a significant contributor to global GHG emissions
- Green Building is the practice of increasing the efficiency of the way buildings use resources: energy, water, and materials while reducing building impacts on human health and the environment during the building's lifecycle, through better siting, design, construction, operation, maintenance, and removal

Effective green building can lead to:

- *reduced operating costs* by increasing productivity and using less energy and water
- *improved public and occupant health* due to improved indoor air quality, and
- *reduced environmental impacts* by, for example, lessening storm water runoff and the heat island

Management of Water Resource

- Only a few cities are already starting to adapt and plan for the long term

- Areas that need to be looked into include supply and availability of fresh water; concerns over contamination of fresh water supplies due to rising sea level; coastal protection against sea level rise and technologies to provide fresh water at minimum cost

- The most important element of such a system is that tremendous amounts of fresh running water are lost throughout the world. Collecting this water would allow large reservoir to be created

- However such system do have serious ecological consequences and could be disastrous for biodiversity if applied on a large scale. This issue must be carefully taken into account

Reforestation and Conservation

There are two necessary topics which need immediate attention and action:

- **Stopping deforestation** across the world (especially in key hot spots areas such as Brazil, Papua New Guinea, Indonesia, Malaysia and New Caledonia)

- **Planting trees** on a large scale. Costa Rica has already taken the lead in this field with its massive reforestation programme. In 2008 about 5 million trees were planted and 7 million are targeted for 2009. Such initiatives must be followed by other larger countries if they are to be effective on a global level

- However, there is no point in reforesting if on the other hand forests continue to deplete at alarming rates. It is essential that actions are taken to address this devastating problem both for climate change and biodiversity issues

- A few projects are underway to develop new technologies that would allow trees to be planted fast and effectively on a large scale, including aerial reforestation

Earth Engineering

- Earth engineering is a new, emerging field which aims to mitigate the worst-case outcomes of climate change

- Earth engineering looks into the adoption of drastic technologies which aim to rapidly alter or modify the climate system

- These are projects which have the potential to affect the entire planet

Such projects include:

- *Covering ice caps with a protective layer which could significantly delay the melting of polar ice*
- *Technological capture of atmospheric CO_2 carbon sequestration*
- *Modifying the world's oceanic currents*
- *Capturing altitude high velocity winds*
- *Reflecting the sun rays with giant mirrors in space*
- *Constantly injecting large amounts of aerosols into the lower atmosphere in order to create a cooling effect (dimming)*

- However, many of these projects are economically unrealistic and would cost trillions of dollars

- Many of such projects are under-researched, and would also have unwanted side effects on ecosystems

- However, if the situation gets catastrophic it is good to have a few options available to rapidly deal with the problem and assure human and the survival of other species.

Case Study: Climate Change and the Business Opportunities that Present themselves

How crisis situations can create the biggest opportunities

Climate change is a fact that affects all aspects of life and every area of business performance, increasing the risk of financial loss and interruption. The United Nations has recently published an alarming report on climate change urging governments, companies, and people alike to take unprecedented action in significantly reducing global carbon emissions by 2030 to avoid catastrophic consequences. Clearly, the need to step up and actually *do* something about it is more intense than ever before. However, amidst these prominent (and potentially devastating) news, tremendous business opportunities are lying ahead.

Understanding Climate Risks

Being committed to addressing exposure to climate risks and building resilience will deliver:

Cost Savings – Businesses can save money by adjusting their maintenance and operation procedures at low (or no) cost to enhance the efficiencies of their equipment. Also, they may save money by modifying equipment and assets to reduce the inputs of their raw materials and boost operational performance. Finally, incorporating climate resilience into the design and specification of assets from the outset can further reduce long-term costs.

Competitive Advantage – Leading companies manage climate risks more effectively and grab opportunities to provide greater value to their customers, reduce costs, and become more efficient. The measures they take (i.e., build internal capacities by collecting data and developing skills) allow them to evolve and adapt.

Business Continuity – Being able to identify risks and change your risk management strategy accordingly is a critical step for business continuity. If everything about the business, from systems and processes to internal policies are geared toward the changing climate, then increased resilience is guaranteed.

Increased Reputation – Shareholders or multi-national corporations are showing a heightened interest in climate resilience. Demonstrating that companies are managing the effects of climate change can improve the reputation of the organization in the eyes of its shareholders. Also, large businesses are collaborating with national governments, NGOs, overseas producers, and development partners to build sustainability and climate resilience into their operations and supply chains. Part of their efforts is to support governments, ecosystems, local communities, and growers. Companies with a human face have been proved to be the receptors of people fondness and enjoy more loyal customers

Seizing the opportunities

Although there are market opportunities for a wide range of sectors, some particular ones seem to benefit slightly more than others. For instance:

Manufacturing – Changes in mean temperatures could provide opportunities for manufacturers of agricultural-supporting products (i.e., pesticides and fertilizers), as well as manufacturers of products that support industry (i.e., specialist equipment to combat water scarcity), and health care.

Insurance – Insurance related to climate (i.e., forest and crop protection) is a rapidly growing global market and an effective tactic for organizations to spread risks and enable fast access to liquidity after a disaster.

Financial services – Developing climate resilient financing mechanisms will open the road for increased access to project finance, bond finance, equity finance, and commercial banks in developing countries. There is also a significant number of potential investment opportunities on the horizon (i.e., precipitation changes will demand the development of technologies and services in the areas of irrigation, flood, and water treatment) so people can cope with the impacts of climate change on the environment.

Construction – Things like flood and overheating risks will challenge the sector and, at the same time, give opportunities to incorporate climate resilience into new developments. The environmental performance of buildings will need to be more efficient, with zero carbon and green infrastructure. The cities will have to be re-designed and engineered so that they can handle climate changes and extreme weather conditions. There will also be lots of opportunities for building maintenance contracts as changes in mean temperature are probably going to boost the carbonization of concrete and steel corrosion, among others. This applies to everything from houses to roads and railways. Of course, this also means that engineering and architectural companies will be major players too.

X. MISUNDERSTANDINGS OF CLIMATE CHANGE

Common Misconceptions About Climate Change

"If we look at past records, we can see that our climate has been constantly changing over time, therefore what we observe today is not alarming"

Indeed our climate has been changing in the geological history of our planet but never in recorded history have we seen such a rapid shift in GHG concentrations and surface temperatures. It is also important to note that previous sudden shifts in atmospheric conditions have resulted in massive extinctions of species which, as a species, we should be more concerned about

"The temperature has only increased by three-quarters of a degree Celsius in the last century, which is not alarming"

This is actually a very large and sudden increase in temperature over such a small period of time. If the climate continues to warm at the present rate (which is predicted to increase), we will undoubtedly face unprecedented disruptions worldwide by the end of the century

"The global sea level has already increased by a few centimeters over the past century which has not caused major troubles so far, why would a few more centimeters or even meters worry us?"

The sea level rise over the past century has been consistently increasing and is showing signs of significantly accelerating. This is a great concern to global security as 80% of our cities and world population is on the coast lines. A sea level rise of even half a meter will cause major troubles worldwide and will displace millions of people. We have seen this happen when a cyclone or a local flood temporally displaces a few thousands (sometime millions) of people. Now imagine what will happen when this will occur on a global scale and permanently! Shouldn't we be more concerned?

"Dealing with Climate Change will be too costly for nations economies, so we cannot do anything about it"

All economist who have seriously analyzed the question come to the same conclusion that the cost of doing nothing in terms of economic and social impacts will far exceed the cost of mitigation measures. The Stern report is a comprehensive analysis on the economics of climate change and provides guidelines to tackle the problem before it is too late and too costly

"Water is abundant on this planet, Climate Change will never be able to deplete our water supplies!"

It is true that we live on a Planet which is 75% water; it is even called "the blue Planet". However, most of the water is stored in oceans and is in a salty form not suitable for human consumption. Fresh water supplies are mostly stored in lakes and underground supplies. However, most of these supplies are already starting to diminish very fast. Climate change will have different impacts in different parts of the world, some parts will drastically become drier while others will experience much more rain (which means more runoff). Both will lead to a loss of available fresh water storage. While rich countries will be able to pay for expensive water treatment plants (desalinisation, ozonation…), poorer countries will be severely affected by droughts and depletion of fresh water supplies which could lead to great instabilities.

"It is still not well established whether or not current observed Global Warming is due to human activity or part of Earths natural variability"

This is not true. The fact that natural variability could explain today's global warming has been ruled out. All known possible causes that have caused the climate to shift in the past cannot explain today's tendency. Powerful computer simulations have been run to support such statements. Today's global warming can only be accounted for when GHG emissions from the burning of fossil fuels since the industrial revolution are taken into account. There is no more doubt that today's climate change is due to human influence

"The long term climate change records suggest that we are on the verge of an ice age so it will offset current global warming"

Yes, it is true that it is likely that we are on the verge of an ice age, However, one must understand that we are here talking in terms of geological time frame. Imminent in geological terms means within the next 100,000 years. Human induced global warming is however a present threat that our generation and the ones to come in the next centuries will have to deal with

"Climate change doesn't concern our generation. Future generations will have to deal with it. Why bother, it won't happen during our life time!"

This statement is wrong for two reasons. First, climate change is happening at present and noticeable changes are beginning to surface. Within, decades serious impacts will start to be felt. Not doing anything about it and giving the responsibility to future generations is an egoistic approach. Secondly, it will be too late for future generations to tackle the source of the problem, their only alternative will be adaptation. Our generation is the one that must do something about climate change while the problem can still be solved

Global warming will affect biodiversity but shouldn't affect us humans"

Such statements are wrong. First climate change will affect human civilizations very significantly in a number of ways which are highlighted in the impacts sections of this booklet. Secondly, the human species are highly dependant on the survival of other species which affect our health and food supplies. A collapse in ecosystems and biodiversity will be a great threat for human survival. There is also a lot to be lost in terms of knowledge. Species which have evolved over hundreds of millions of years are marvels of adaptation to specific environments and we have a lot to learn from them

Maybe for our own survival! Most of our medicines come from plants and animals. Losing biodiversity will also mean the loss of molecules which could hold remedies for a variety of diseases such as cancer, AIDS and other life threatening diseases... It is highly unlikely that humans would be able to artificially synthesize such complex molecules without studies and samples from real organisms. Biodiversity is of great value to us humans and should be preserved at all costs. One must remember that extinction of species cannot be reversed and is forever!

"Our Planet can take whatever we do to it. Earth will absorb the excess carbon dioxide that we put into the atmosphere, so we do not need to worry about it! "

It is true that the earth has a certain buffering capacity. For instance, the oceans have been absorbing much of the excess carbon dioxide that we have emitted into the atmosphere since the industrial revolution. However, there is a downside to it.

By absorbing excess CO_2 the oceans are turning acidic which is starting to have serious impacts on marine life and the well being of our oceans. It is estimated that within half a century the capacity of ecosystems to uptake CO_2 will be exceeded which will even worsen the rate of warming. Earth is a fragile planet and our actions have severe and irreversible impacts on its environment. If we keep on pushing ecosystems to the limits they will end up collapsing along with our food supplies and complex interactions which makes this planet hospitable to us humans

"Governments are taking care of the problem, so as a corporate entity or as an individual I do not need to do anything. Even if I do something at the individual level it will not change anything!"

This is a wrong approach. First, governments are doing very little to seriously tackle the problem at present. Much stronger measures are required and most importantly a mutual action by world nations is still needed in order to change things. Secondly, corporations and individuals play a very important role in emitting GHG. Small contributions, if applied on a large scale will make a significant difference. The more individuals are involved the more the message for action will spread and be widely adapted like a snowball effect. Corporations and individuals must not wait for governments to implement regulations but take matters into their own hands.

- There is not a single day without Climate Change related issues in the press

- Evidence of a rapidly changing planet is all around, one has just to make the connection

- There are very few people (with credible academic backgrounds) who argue that present climate change is not caused by human activity. The vast majority of leading scientists from various disciplines all over the world have no more doubt about this...as concluded in the recent IPCC report compiling the works of over 2500 scientists.

"Fighting climate change by adopting clean technologies and practices is too expensive. It is not achievable!"

This is not true. At present the experience shows that installation of sustainable energies and practices makes a lot of sense in terms of economical profit and long term investment. For instance, building a sustainable building with low emissions might cost about 3-4% more, but the return period is only about 4-5 years. Savings on energy bills will allow the developer to save money in the long term. The same principle applies at all levels (including individuals and government): the extra cost of initial investment is annulled by long term benefits, not to forget the benefit of reducing GHG emissions

"Global Warming! Great, I live in a cold country at high latitude so that means our winters will be less cold and shorter! "

For someone who doesn't like the cold weather, indeed global warming might sound like a pleasant idea. Unfortunately, benefits from global warming will be far exceeded by negative effects. One thing that we are sure of is that as the climate will warm it will also become more uncertain and erratic. Extreme events such as ice storms, heavy snow fall, rains, floods and so on will be more frequent. This and other impacts will create great instabilities worldwide for societies and nations economies. For the past centuries we have lived in a climate relatively stable which may soon change due to climate change.

"I find the predictions about Climate Change a bit confusing and inconsistent. Most people are talking about a Global Warming but some recent movies and documentaries suggest a Global rapid cooling!"

Those people may be referring, among others, to the movie *"The Day after Tomorrow"* suggesting an extremely rapid shift from a warm world to an ice age. This movie despite being fictional in the rapidity at which such event could occur is based on real scientific facts. Most of our climate is governed by oceanic currents which transport heat from the poles to the equator. There is scientific evidence that suggest that if the planet continues to warm up at the poles, the variation in salinity in the arctic as a consequence of ice melt will result in a slowing down of the oceanic currents which are initiated in this area through normally very salty waters. The fear is that if this mechanisms, acting as a giant pump, slows down to the point of stopping it will indeed result in global and rapid perturbation of the Climate System in very short periods of time (decades to a century). This is not fiction, but a possible scenario. Scientist monitoring the phenomenon are noticing a slowdown of the oceanic circulation but the situation seems to remain stable at present. How long are we willing to gamble?

"There are too many uncertainties in the predictions of Climate Change, why should we take it seriously?"

It is true that there are many uncertainties in the predictions on how our climate will react to change. However, there are also many aspects of the climate system which we do know for sure, based on past and present observations and our general understanding of the physical principals driving the climate on earth. There are no doubts that the predictions stated will occur; what we do not know is the exact extent and severity of occurrence. In many instances predictions have been made within a range which they are most likely to occur in. However, for many predictions there is great fear that they are actually underestimating the threat in many aspects as suggested by several new reports from leading science organizations. In all cases ignoring the threat is not an acceptable option; in the fear that it will occur preventive measures and preparation is an ethical and moral obligation

"In the past we have been able to find a solution to resolve severe environmental problems such as the Ozone hole, when time comes critical we will find something to save the day!"

It is true that the ozone crisis has partly been resolved by banning certain types of molecules in products. However, first it is still an ongoing problem secondly there is not one simple solution to tackle climate change and the more we wait, the more difficult it will be to fix the problem (we might even pass the point of no return within the century if GHG concentration pass 650 ppm). Tackling climate change requires immediate action and global collaboration

"Extreme weather events have been occurring for ever, impacts of climate change on the weather will not bring anything new. Therefore there is nothing to fear"

Climate change is expected to make our weather more unpredictable and the occurrence of extreme events such as heavy rain, droughts, tropical storms are expected to increase significantly. The severity of these events is also expected to increase. Stronger and more frequent events will put a lot of pressure on societies and numerous lives will be at risk. Such events will also create major economical and social instabilities by displacing millions of people as weather refugees. There is a lot to worry about as our civilizations are highly vulnerable to weather extremes

"I live in Singapore, and for now I don't really see strong evidence around me that indicate that we are in trouble. Climate change is happening overseas and does not concern us"

Such argument is wrong. First even in places like Singapore impacts of change are starting to be felt and will worsen in the decades to come (e.g. sea level rise, economical difficulties). One must understand that climate change is a global problem and that no one will be spared. We live on a global planet in which what happens in one part of the world has direct repercussions in other parts of the world as well (e.g. food prices). If one country becomes unstable due to catastrophic water shortages or loss of food supplies this will create tensions between neighbouring nations. If a country loses land area and is affected by natural disasters to the extent that they can no longer sustain themselves, refugees will have to go somewhere! There is simply no alternative to solving the climate crisis then global collaboration in solving a planetary emergency. Everyone must start doing their part at the individual, corporate and governmental levels

"All predictions on future climate change are based on computer models. I don't think I can trust computers!"

In the early days a lot of data was collected and analyzed by hand which enabled early scientists such as Roger Revelle to make predictions on the probable evolution of our climate. We have gone a long way since then and extremely powerful computers now enable scientists to run simulations that would take centuries to do by hand. These simulations are simply a representation of our climate system based on the knowledge that we have of it. They follow the same rules of physics as the real world and therefore the predictions give a good representation of what is likely to happen over time. Similar models are put to the test everyday in meteorological predictions. Computers are just a tool to run simulations. Strong scientific understanding and knowledge is behind the models

"Even if we take active actions to tackle Climate Change, the climate will still continue to warm up. So why should we do anything about it anyway?"

It is true that due to the life time of GHG in the atmosphere and the inertia of the climate system, no matter what we do temperatures will continue to rise at a rate of a at least 0.1 °C per decade. However, doing nothing would be a terrible mistake as there is still time to limit the damages. There is still time to stabilize the GHG in the atmosphere to acceptable levels, however time is running out!

"Fossil fuels is a sustainable resource there are still reserves for centuries and there is no reasons to stop exploiting it"

This is wrong. First burning fossil fuels at the current rates is probably the most inefficient way of utilizing the resource (e.g. making of plastics). Secondly fossil fuels have originated from the decay of organic matter over millions of years and the supplies are rapidly running out. The global reserves of petroleum will be utilized within 30 years. Coal and natural gas have however supplies for probably another century. One thing for sure is that fossil fuel is not a renewable resource and it is not sustainable as, in the process of burning it, we are releasing carbon into the atmosphere which has been stored in the Earth crust for millions of years. It is now established that this is by far the main cause of present global warming and that if we continue to do so at the current rate, the future of our climate and environment as we know it today is heading towards catastrophic scenarios

Climate Change definitions:

Albedo
The fraction of solar radiation reflected by a surface or object, often expressed as a percentage. Snow covered surfaces have a high albedo; the albedo of soils ranges from high to low; vegetation covered surfaces and oceans have a low albedo. The earth's albedo varies mainly through varying cloudiness, snow, ice, leaf area and land cover changes

Climate projection
A projection of the response of the climate system to emission or concentration scenarios of greenhouse gases and aerosols, or radiative forcing scenarios, often based upon simulations by climate models. Climate projections are distinguished from climate predictions in order to emphasise that climate projections depend upon the emission/ concentration/ radiative forcing scenario used, which are based on assumptions, concerning, e.g., future socio-economic and technological developments, that may or may not be realised, and are therefore subject to substantial uncertainty

Climate system
The climate system is the highly complex system consisting of five major components: the atmosphere, the hydrosphere, the cryosphere, the land surface and the biosphere, and the interactions between them. The climate system evolves in time under the influence of its own internal dynamics and because of external forcings, such as volcanic eruptions, solar variations and human-induced forcings such as the changing composition of the atmosphere and land-use change

Climate variability
Climate variability refers to variations in the mean state and other statistics (such as standard deviations, the occurrence of extremes, etc.) of the climate on all temporal and spatial scales beyond that of individual weather events. Variability may be due to natural internal processes within the climate system (internal variability), or to variations in natural or anthropogenic external forcing (external variability)

Emission scenario
A plausible representation of the future development of emissions of substances that are potentially radiatively active (e.g. greenhouse gases, aerosols), based on a coherent and internally consistent set of assumptions about driving forces (such as demographic and socio-economic development, technological change) and their key relationships. Concentration scenarios, derived from emission scenarios, are used as input into a climate model to compute climate projections

Energy balance
Averaged over the globe and over longer time periods, the energy budget of the climate system must be in balance. Because the climate system derives all its energy from the sun, this balance implies that, globally, the amount of incoming solar radiation must on average be equal to the sum of the outgoing reflected solar radiation and the outgoing infrared radiation emitted by the climate system. A perturbation of this global radiation balance, be it human induced or natural, is called radiative forcing

Extreme weather event
An extreme weather event is an event that is rare within its statistical reference distribution at a particular place. Definitions of 'rare' vary, but an extreme weather event would normally be as rare as or rarer than the 10th or 90th percentile. By definition, the characteristics of what is called extreme weather may vary from place to place. An extreme climate event is an average of a number of weather events over a certain period of time, an average which is itself extreme (e.g. rainfall over a season)

Global surface temperature
The global surface temperature is the area-weighted global average of (i) the sea-surface temperature over the oceans (i.e. the subsurface bulk temperature in the first few meters of the ocean), and (ii) the surface-air temperature over land at 1.5 m above the ground

Global Warming Potential (GWP)
An index, describing the radiative characteristics of well mixed greenhouse gases, that represents the combined effect of the differing times these gases remain in the atmosphere and their relative effectiveness in absorbing outgoing infrared radiation. This index approximates the time-integrated warming effect of a unit mass of a given greenhouse gas in today's atmosphere, relative to that of carbon dioxide

Infrared radiation
Radiation emitted by the earth's surface, the atmosphere and the clouds. It is also known as terrestrial or long wave radiation. Infrared radiation has a distinctive range of wavelengths ('spectrum') longer than the wavelength of the red colour in the visible part of the spectrum. The spectrum of infrared radiation is practically distinct from that of solar or short wave radiation because of the difference in temperature between the sun and the earth-atmosphere system

Land-use change
A change in the use or management of land by humans, which may lead to a change in land cover. Land cover and land-use change may have an impact on the albedo, evapotranspiration, sources and sinks of greenhouse gases, or other properties of the climate system

Milankovitch cycles
Milankovich cycles are cycles in the earth's orbit that influence the amount of solar radiation striking different parts of the earth at different times of year. They are named after a Serbian mathematician, Milutin Milankovitch, who explained how these orbital cycles cause the advance and retreat of the polar ice caps

Ocean conveyor belt
The theoretical route by which water circulates around the entire global ocean, driven by wind and the thermohaline circulation

Radiative forcing
Radiative forcing is the change in the net vertical irradiance (expressed in Watts per square metre: Wm^{-2}) at the tropopause due to an internal change or a change in the external forcing of the climate system, such as, for example, a change in the concentration of carbon dioxide or the output of the sun. Usually radiative forcing is computed after allowing for stratospheric temperatures to readjust to radiative equilibrium, but with all tropospheric properties held fixed at their unperturbed values. Radiative forcing is called instantaneous if no change in stratospheric temperature is accounted for

Sink
Any process, activity or mechanism which removes a greenhouse gas, an aerosol or a precursor of a greenhouse gas or aerosol from the atmosphere.

Solar radiation
Radiation emitted by the sun. It is also referred to as short wave radiation. Solar radiation has a distinctive range of wavelengths (spectrum) determined by the temperature of the sun. See: Infrared radiation.

Thermal expansion
In connection with sea level, this refers to the increase in volume (and decrease in density) that results from warming water. A warming of the ocean leads to an expansion of the ocean volume and hence an increase in sea level

Thermohaline circulation
Large-scale density-driven circulation in the ocean, caused by differences in temperature and salinity. In the North Atlantic the thermohaline circulation consists of warm surface water flowing northward and cold deep water flowing southward, resulting in a net poleward transport of heat. The surface water sinks in highly restricted sinking regions located in high latitudes

Acronyms and abbreviations

CF4 perfluoromethane
CFC Chlorofluorocarbons
CFC-11 trichlorofluoromethane
CH4 methane
CO2 carbon dioxide
ENSO El Niño - Southern Oscillation
FCCC UN Framework Convention on Climate Change
GCM General Circulation Model
GCOS Global Climate Observing System
GHG Greenhouse Gas
GSN GCOS Surface Network
GtC Gigatonnes of Carbon
HCFC hydrochlorofluorocarbons
HFC hydrofluorocarbons
ICSU International Council for Science
IEA International Energy Agency
IGBP International Geosphere-Biosphere Programme
IOC Intergovernmental Oceanographic Commission
IPCC Intergovernmental Panel on Climate Change
MSLP Mean sea-level pressure
N2O nitrous oxide
NAO North Atlantic Oscillation
OECD Organisation for Economic Cooperation and Development
OECD Organisation for Economic Cooperation and Development
SOI Southern Oscillation Index
SST Sea-surface temperature
UN United Nations
UNCED United Nations Conference on Environment and Development
UNEP United Nations Environment Programme
UNESCO United Nations Educational, Scientific and Cultural Organization
UNFCCC United Nations Framework Convention on Climate Change
WCP World Climate Programme
WCRP World Climate Research Programme
WMO World Meteorological Organization
WSSD World Summit on Sustainable Development

Climate Change Quotes

"Scientific evidence for warming of the Climate System is unequivocal." **Intergovernmental Panel on Climate Change**

"There is still time to avoid the worst impacts of climate change, if we take strong action now." **Sir Nicholas Stern**

"The debate is over! There's no longer any debate in the scientific community about this. But the political systems around the world have held this at arm's length because it's an inconvenient truth" **Al Gore**

"We are in an extremely precarious and urgent situation that compels immediate action" **Prof. David Karoly**

"For my generation, coming of age at the height of the Cold War, fear of nuclear winter seemed the leading existential threat on the horizon. But the danger posed by war to all humanity—and to our planet—is at least matched by climate change." **Ban Ki-moon**

"Ignoring climate change will be the most costly of all possible choices, for us and our children"
Peter Ewins

"There is no doubt in my mind that this is the greatest problem confronting mankind at this time and that it has reached the level of a state of emergency" **Prof David de Kretser**

"We are playing Russian roulette with features of the planet's atmosphere that will profoundly impact generations to come. How long are we willing to gamble?" **David Suzuki**

"First, I worry about climate change. It's the only thing that I believe has the power to fundamentally end the march of civilization as we know it, and make a lot of the other efforts that we're making irrelevant and impossible." **Bill Clinton**

"The choice to 'do nothing' in response to the mounting evidence is actually a choice to continue and even accelerate the reckless environmental destruction that is creating the catastrophe at hand"
Albert Arnold Gore

"We need to push ourselves to make as many reductions as possible in our own energy use first...and that takes time. But we must do this quickly... the climate will not wait for us.' **Rupert Murdoch**

I want to testify today about what I believe is a planetary emergency - a crisis that threatens the survival of our civilization and the habitability of the Earth" **Al Gore**

"Global warming is too serious for the world any longer to ignore its danger or split into opposing factions on it" **Tony Blair**

"The Greenland ice sheet is likely to be eliminated [within 50 years] unless much more substantial reductions in emissions are made than those envisaged [and will] probably be irreversible, this side of a new ice age." **Jonathan Gregory, climatologist**

"Climate change is the most severe problem that we are facing today, more serious even than the threat of terrorism." - **David King, UK government chief scientific adviser**

"The absolutely best case scenario - which in my opinion is unrealistic - with the minimum expected climate change... we end up with an estimate of 9% [of all species] facing extinction." **Chris Thomas, ecologist at the University of Leeds**

"The warnings about global warming have been extremely clear for a long time. We are facing a global climate crisis. It is deepening. We are entering a period of consequences" **Al Gore**

Of the more than 29,000 observational data series, from 75 studies, that show significant change in many physical and biological systems, more than 89% are consistent with the direction of change expected as a response to warming..." **IPCC**

"Anthropogenic warming could lead to some impacts that are abrupt or irreversible, depending upon the rate and magnitude of the climate change." **IPCC**

Biography

- Alley, R., T. Berntsen, N.L. Bindoff, Z. Chen, A. Chidthaisong, P. Friedlingstein, J. Gregory, G. Hegerl, M. Heimann, B. Hewitson, B. Hoskins, F. Joos, J. Jouzel, V. Kattsov, U. Lohmann, M. Manning, T. Matsuno, M. Molina, N. Nicholls, J. Overpeck, D. Qin, G. Raga, V. Ramaswamy, J. Ren, M. Rusticucci, S. Solomon, R. Somerville, T.F. Stocker, P. Stott, R.J. Stouffer, P. Whetton, R.A. Wood and D. Wratt (2007). Climate change 2007: The physical science basis - summary for policy makers - IPCC WGI fourth assessment report. Geneva, Switzerland, Intergovernmental Panel on Climate Change (IPCC): 21.

- Footitt, A., M. McKenzie, P. Kristensen, A. Leipprand, T. Dworak, R. Wilby, J. Huntington, J. van Minnen, R. Swart, A.Jol and M.M. Hedger (2007). Climate change and water adaptation issues (EEA technical report) . Copenhagen (Denmark), European Environment Agency: 114.

- McGranahan, G., D. Balk and B. Anderson (2007). "The rising tide: assessing the risks of climate change and human settlements in low elevation coastal zones." Environment & Urbanization 19(1): 17-37.

- Alcamo, J., M. Florke and M. Marker (2007). "Future long-term changes in global water resources driven by socio-economic and climatic changes." Hydrological Sciences Journal 52(2): 247-275. 2007

- Brumbelow, K. and A. Georgakakos (2007). "Consideration of climate variability and change in agricultural water resources planning." Journal of Water Resources Planning & Management 133(3): 275-285.

- Schellnhuber, H. J., W. Cramer, N. Nakicenovic, T. Wigley and G. Yohe, Ed. (2006). Avoiding dangerous climate change. Cambridge, UK, Cambridge University Press.

- Stern, N. (2006). Stern review: the economics of climate change. London, Cabinet Office - HM Treasury (UK): 575.

- Hare, B. (2006). Relationship between increases in global mean temperature and impacts on ecosystems, food production, water and socio-economic systems. Avoiding dangerous climate change. H. J. Schellnhuber, W. Cramer, N. Nakicenovic, T. Wigley and G. Yohe. Cambridge, UK, Cambridge University Press: 177-185.

- Thompson, L.G. , E.M. Thompson, H. Brecher, M. Davis, B. Leon, D. Les, P.N. Lin, T. Mashiotta and K. Mountain (2006). "Abrupt tropical climate change: past and present." Proceedings of the National Academy of Sciences 103(28): 10536-10543.

- Leipprand, A. and D. Gerten (2006). "Global effects of doubled atmospheric CO2 content on evapotranspiration, soil moisture and runoff under potential natural vegetation." Hydrological Sciences Journal 51(1): 171-185.

- Pearce, F. (2005). "Cities may be abandoned as salt water invades." New Scientist 186(2495): 9-9.

- Alley, R. B. , P. U. Clark, P. Huybrechts and I. Joughin (2005). "Ice-sheet and sea-level changes." Science 310(5747): 456-461

- Barnett, T.P., J.C. Adam and D.P. Lettenmaier (2005). "Potential impacts of a warming climate on water availability in snow-dominated regions." <u>Nature</u> 438(7066): 303-309.

- Hay, J.E. and N. Mimura (2005). "Sea-level rise: implications for water resources management." <u>Earth and Environmental Sciences</u> 10(4): 717-737.

- Hambrey, M.J. and A. Jurg (2004). <u>Glaciers</u>. Cambridge, UK, Cambridge University Press.

- Gregory, J. M., P. Huybrechts and S.C.B. Raper (2004). "Threatened loss of the Greenland ice-sheet." <u>Nature</u> 428: 616.

- Hitz, S. and J. Smith (2004). "Estimating global impacts from climate change." <u>Global Environmental Change</u> 14(3): 201-218

- Zhang, K., B.C. Douglas and S.P. Leatherman (2004). "Global warming and coastal erosion." <u>Climatic Change</u> 64(1-2): 41-58

- Barnett, J. (2003). "Security and climate change." <u>Global Environmental Change</u> 13(1): 7-17.

- Karl, T.R. and K.E. Trenberth (2003). "Modern global climate change." <u>Science</u> 302(5651): 1719-1723.

- Lambeck, K., T. M. Esat and E.-K. Potter (2002). "Links between climate and sea levels for the past three million years." <u>Nature</u> 419: 199 - 206.

- Beeton, A. M. (2002). "Large freshwater lakes: present state, trends, and future." <u>Environmental Conservation</u> 29(1): 21-38.

- Lal, M., H. Harasawa and K. Takahashi (2002). "Future climate change and its impacts over small island states." <u>Climate Research</u> 19(3): 179-192.

- Hulme, M. and D. Viner (1998). "A Climate Change Scenario for the Tropics." <u>Climatic Change</u> 39(2): 145-176

- Gregory, J. M. and J. Oerlemans (1998). "Simulated future sea level rise due to glacier melt based on regionally and seasonally resolved temperature changes." <u>Nature</u> 391: 474-476.

- Pahl-Wostl, C. (2007). "Transitions towards adaptive management of water facing climate and global change." <u>Water Resources Management</u> 21(1): 49-62.

- Alcamo, J., M. Florke and M. Marker (2007). "Future long-term changes in global water resources driven by socio-economic and climatic changes." <u>Hydrological Sciences Journal</u> 52(2): 247-275.

- Halsnaes, K. and J. Verhagen (2007). "Development based climate change adaptation and mitigation - conceptual issues and lessons learned in studies in developing countries." Mitigation and Adaptation Strategies for Global Change online.

- Brinkuis, H. , S. Schouten, M. E. Collinson, A. Sluijs, J. S. S. Damste, G. R. Dickens, M. Huber, T. M. Cronin, J. Onodera, K. Takahashi, J. P. Bujak, R. Stein, J. van der Burg and J. S. Eldrett (2006). "Episodic fresh surface waters in the Eocene Arctic Ocean." Nature 441(7093): 606.

- Schellnhuber, H. J., W. Cramer, N. Nakicenovic, T. Wigley and G. Yohe, Ed. (2006). Avoiding dangerous climate change. Cambridge, UK, Cambridge University Press.

- Oki, T. and S. Kanae (2006). "Global hydrological cycles and world water resources." Science 313(5790): 1068-1072.

- Sauerborn, R. and L.R. Valerie, Ed. (2007). Global environmental change and infectious diseases: impacts and adaptation strategies, Springer.

- Milly, P.C.D. (2007). Global warming and water availability: the "big picture". 21st Conference on Hydrology - The 87th American Meteorlogical Society (AMS) Annual Meeting. San Antonio, TX, American Meteorlogical Society.

- Hare, B. (2006). Relationship between increases in global mean temperature and impacts on ecosystems, food production, water and socio-economic systems. Avoiding dangerous climate change. H. J. Schellnhuber, W. Cramer, N. Nakicenovic, T. Wigley and G. Yohe. Cambridge, UK, Cambridge University Press: 177-185.

www.ingramcontent.com/pod-product-compliance
Lightning Source LLC
Chambersburg PA
CBHW031415210526
45464CB00005B/1890